Inspection Guidelines for Pressure Vessels and Piping

Volume 1: New Fabrication

MTI Publication No. 40

Ronald J. Twigg
Glencor Engineering Limited
Nepean, Ontario, Canada

Published for the Materials Technology Institute of the Chemical Process Industries, Inc. by NACE International

Library of Congress Catalog Card Number: 93-85984
ISBN: 1-877914-64-9
Printed in the United States of America

Published for Materials Technology Institute of the Chemical Process Industries, Inc.

By NACE International
P.O. Box 218340
Houston, TX 77218-8340

Information on availability of this publication can be obtained from NACE International

About MTI

The Materials Technology Institute of the Chemical Process Industries, Inc., (MTI) is a unique, cooperative research and development organization representing private industry. Its objective is to conduct generic, nonproprietary studies of a practical nature on the selection, design, fabrication, testing, inspection, and performance of materials and equipment used in the process industries.

Through membership in MTI, companies can solve nonproprietary problems of major concern to the process industries, leverage research and development dollars by participating in the direction and results of MTI studies, capitalize on the expertise of member company representatives, and learn about MTI's accomplishments in a timely fashion.

More information about membership in MTI can be obtained from the Executive Director. MTI members receive copies of MTI publications as part of their membership.

The Materials Technology Institute
of the Chemical Process Industries, Inc.
1215 Fern Ridge Parkway, Suite 116
St. Louis, MO 63141-4401
Tel: 314/576-7712
Fax: 314/576-6078

Contents

CHAPTER 7

CHAPTER 8

Acknowledgments

The author would like to thank members of the MTI Resource Group for Project No. 62 for helpful discussions and suggestions during the preparation of this document; in particular, Executive Director of MTI, A.S. (Bert) Krisher; R.W. (Bud) Ross, formerly MTI Technical Consultant and now of R.W. Ross & Associates; Robert E. Smallwood of Cytec Industries Inc., formerly the chemical division of American Cyanamid Company; F.Galen Hodge of Haynes International Inc; Eugene Redko of FMC Corporation; Nick Monsour, formerly of Nova Petrochemical Inc.; and Mati Tiivel of ICI Canada Inc., for their review, comments, and helpful suggestions. Thanks are extended as well to a number of other reviewers who sent in helpful comments but neglected to identify themselves.

Special thanks are extended to American Cyanamid Company for permission to use the "Nuts and Bolts" publication in this document as well as to a number of other companies who supplied data sheets, sketches, and other information used as examples but who wished to remain anonymous.

The editorial, publication, and design assistance of Peter H. Koops, Kenneth J. Martinez, Todd R. Overton, and Aunee Pierce at NACE International is gratefully acknowledged.

The author would like to express his appreciation to a number of "old time" inspectors who have unknowingly contributed to this publication by passing on invaluable information over the years and to a number of shop inspectors for discussions on current practices and shortcomings.

The front cover photo is provided courtesy of Haynes International Inc.; the photo on the back cover is courtesy of Rolled Alloys.

Executive Summary

Obtaining quality processing equipment has become increasingly difficult over the last number of years, partially because of the retirement of craftsmen inspectors and the movement towards reducing inspection as a cost savings means. However, many companies, realizing that inspection is necessary to ensure a quality product, are attempting to train new inspectors without the benefit of experienced inspectors' advice.

This guideline gives procedures that may be used to ensure quality of new shop fabricated pressure vessels and piping. It outlines a detailed program for critical vessels. Less critical equipment may be inspected in accordance with the program as well, simply by deleting nonapplicable inspection details and modifying check sheets accordingly. Emphasis is placed on what, when, and how to check various items to obtain quality fabrication.

Brief descriptions of the most common nondestructive testing methods used to assess the quality of vessels and piping are also given. Additional technical data and details may be found in the appendixes. This information is intended to assist the inspector by providing common reference material.

CHAPTER 1

Introduction

Background

These guidelines are designed to ensure the quality of new pressure vessels and piping for process industries. They are intended for use by inspectors of the purchasing company or third party inspectors hired by the purchaser to monitor its contracts. To ensure compliance with contract documents, emphasis is placed rather on verifying the efforts of fabricator inspectors than on the actual tests performed. This does not preclude the purchaser's representative from carrying out such tests as may be required to ensure the quality of the component. However, it must be emphasized that the purchaser, although performing acceptance inspections, is not absolving the fabricator of responsibility for defective products.

As the guidelines provide quality requirements for pressure vessels and piping, American Society of Mechanical Engineer's (ASME)[(1)] "Boiler and Pressure Vessel Codes (B&PV)," Appendix 10,[1] is a most useful resource for identifying areas or items requiring monitoring or verification testing. These guidelines follow the recommended policies and procedures established by ASME for ensuring the quality of pressure components, but it also includes a number of concepts based on shop inspection experience to ensure that the purchaser obtains the desired level of quality.

In order to have a clear basis for, and understanding of the inspection function, the scope and purpose of the work to be performed must be fully outlined. Such an outline would include not only the type of work that is to be accomplished, but also state how and by whom the work is to be completed. This detailed statement could prevent future conflicts over any perceived jurisdictional disputes.

Thorough inspection of new pressure vessel and piping systems starts long before the first order is placed with the fabricating shop. The inspection function requires that a high degree of communication be established and maintained between all parties, including: Project Engineering, Design Engineering, Process Engineering, Maintenance Engineering, Purchasing, Specialist Groups, and any consultants that the purchaser may use on the project. It is costly to bring requirements to the design, or even procurement stage, and then find out that the materials are inadequate, regulatory agencies will not approve the construction, and/or fabricators cannot build the component economically using conventional techniques. Continuing communication among the groups listed can reduce the possibility of such situations occurring.

Discussions would also identify equipment that is considered critical (requiring considerable inspection effort) and noncritical (requiring less inspection effort). For example, pressure vessels and piping containing hydrocarbons or other flammable, toxic, or unstable compounds would generally be identified as critical, while water systems or those containing innocuous substances might be considered noncritical. Care will have to be exercised in the criticality analysis because failure of a simple cooling water component could result in loss of coolant in an entire plant, causing an extensive

[(1)]American Society of Mechanical Engineers, New York, NY.

shutdown. Therefore, the active involvement of representatives from all plant functions of the project team is important.

As applied to vessels and piping built of many alloys and for a variety of processes, this information is general in nature and does not attempt to address the fabrication of specific alloys. In addition, check lists and other details on what is to be examined are intended to outline as much information as possible, yet remain within a general guideline. Consequently, all items may not apply to every specific contract; therefore, it is up to the inspector to decide which items are applicable and which may be deleted or added to suit the individual job requirements.

CHAPTER 2

Purchaser and Manufacturer Responsibilities

Customer's Responsibilities

Purchasers of pressure equipment for the process industries have long recognized that a considerable inspection effort is needed to ensure that equipment meets the quality requirements outlined in purchaser specifications and supporting documents. Generally, they also accept that quality cannot be inspected *into* a component; it must be engineered and/or designed in *before* fabrication starts. Yet, these same individuals may purchase pressure components strictly on a "lowest price" criterion without giving any thought to the technical capabilities of the supplying facility. When problems arise, the purchaser often faces considerable losses in terms of time, effort, and expenses helping the fabricator to supply components that will meet the purchaser specifications.

Many cost overruns, poor product quality, and general ill will could be avoided if purchasers and suppliers would each outline their responsibilities and adhere to a few basic common-sense fundamentals related to procurement.

The customer knows where the component is to be used, the process parameters, and the consequences of a failure; therefore, *it is the customer who must decide upon the level of quality required.* The customer must be completely assured that the supplier is capable of producing satisfactory products on a continuing basis.

The customer's responsibilities generally include the following:

1. Evaluating and selecting of suppliers based on technical ability and prior history of supplying quality products. Bidders lists are prepared following evaluation visits, usually by inspectors, and contracts are issued only to qualified bidders. The bidders list is updated periodically to remove poor performers and to add new ones who have proved their technical competence.
2. Evaluating of fabricators' quality manuals, programs, procedures, and work methods to ensure that they are adequate to meet the contract requirements.
3. Preparing of clear and concise tender documents, purchase orders, drawings, etc., outlining, in detail, all of the requirements that are to be met by the fabricator. Subcontractor control expectations are outlined as well.
4. Outlining of traceability requirements prior to starting work to ensure that a proper record is maintained.
5. Providing a list of the documents to be submitted as part of the contract.
6. Providing a list of regulatory requirements that apply to the contract.
7. Carrying out quality surveillance, inspection, or audits, as necessary, to ensure that the contractor is meeting the specified requirements. Subcontractor visits may also be required.
8. Releasing completed components promptly to avoid shipping delays, unnecessary storage costs, or overcrowding of the contractor's facilities.

9. Informing the contractor promptly about any unacceptable components, processes, or procedures.
10. Assisting in the timely resolution of problems.
11. Paying the contractor's invoices promptly for work completed.

Supplier's (Contractor's) Responsibilities

The supplier's responsibilities generally encompass the following:

1. Satisfying the requirements of the contract and specifications given. This includes a thorough review of all contract documents to ensure that the requirements are understood prior to starting the work.
2. Preparing and issuing quality drawings, specifications, procedures, inspection and test plans, and any other materials required by the contract. Utilizing the latest revision of such documents and replacing obsolete ones.
3. Providing adequate evidence through inspection and documentation that the work is of acceptable quality.
4. Periodically calibrating all testing instruments, gauges, and measuring devices to ensure that they are acceptable to a recognized standard.
5. Providing the customer's representative and regulatory inspectors access to the work area. Often, this access includes the provision of office or desk space for work and a lockable storage space for documents.
6. Periodically reviewing and/or auditing the quality program to ensure that it continues to meet the contract requirements.
7. Ensuring that the proper component identification is maintained.
8. Providing adequate protection for components stored prior to shipment.
9. Promptly informing the customer of any problems affecting the contract, including potential labor difficulties.
10. Providing adequate training and inspection instruments for inspectors.
11. Promptly correcting defects identified by inspection, including deviations from approved manufacturing, quality, or testing procedures.
12. Ensuring that all subcontractors meet the quality requirements of the contract.
13. Providing adequate storage and "quarantine" areas to hold incoming material or completed goods.
14. Promptly completing all fabrication documentation, ensuring that all required testing has been witnessed and accepted by authorized inspectors.
15. Paying all subcontractors promptly to avoid the possibility of litigation that would cause shipping delays.

CHAPTER 3

Pre-order Activities

Resource Team Responsibilities

Pre-order activities involve many disciplines that vary from plant to plant depending on job descriptions, company policies, and assigned responsibilities. For example, some plants have materials engineers, welding engineers, and fabrication specialists in a central engineering group. Others place these individuals in inspection or maintenance departments, or attach them to various operating units. Therefore, responsibilities for various pre-order activities will be established by company policy. An effective technical resource team includes engineering personnel from various areas, including inspection, purchasing, and process, as a minimum. In general, the team's activities include:

1. Initial meetings to outline the type of equipment to be purchased;
2. An outline of the process conditions to be experienced;
3. Selection of construction materials and fabricating techniques based on process conditions;
4. Outline of proposed changes to existing equipment that may be required;
5. List of regulatory agency requirements, discussion with the local representative and/or an insurance agency representative may be necessary;
6. Preparation and review of drawings by all concerned personnel to ensure compliance with codes, standards, and company policy;
7. A detailed outline of inspection requirements;
8. Preparation and critical review of specifications by various technical personnel; and
9. Preparation and review of the purchase order by the technical personnel assembled as the resource team.

Bid Packages

Upon completion of the pre-order activities, a bid package is sent to various suppliers or fabricators with an invitation to bid. This bid package contains, as a minimum:

1. At least three copies of the drawings so that the fabricator may use one as a working copy for a bid, return a marked copy with the bid, and retain a copy for reference;
2. The specification to be used;
3. A list of required materials;
4. Instructions on how to submit a bid;
5. Inspection requirements, which also indicate what documentation is to be retained;
6. Shipping and handling instructions;
7. Protection and preservation requirements, if applicable;
8. Right of access for purchaser inspectors; and
9. Regulatory agency requirements.

CHAPTER 4

Supplier/Fabricator Evaluation

Evaluation Requirements

Supplier evaluation is a critical and necessary function to ensure the receipt of properly built pressure equipment and piping. An evaluation assesses the capabilities of a supplier to both meet the desired quality requirements and to supply equipment on time at the most economical prices. A list could be prepared of those firms meeting an established minimum standard and requests for bids could then be sent only to those companies listed. A yearly updating of the lists would remove poor performers as well as add new companies.

Evaluation generally requires at least one plant visit to examine the shop and the various fabrication processes. Frequently, the fabricator is asked to submit a quality manual as well as fabricating and testing procedures for evaluation prior to the shop visit. Obviously, if the shop is using unacceptable procedures, a plant visit is a waste of time.

Items to be considered during a site visit may include: equipment and layout, personnel and management policy, welding, testing methods, gauging, and other equipment.

Equipment and Layout Considerations

The following equipment and layout considerations should be taken into account.

1. Does the fabricator have the necessary machinery and equipment to handle the work or will subcontracting a portion be necessary? Ask for an equipment list.
2. Is the equipment well maintained?
3. Is the shop physically large enough to handle the order? Does it have sufficient lighting?
4. Are shop doors large enough to permit access to the component by any necessary equipment? Are they also large enough to allow the component to be removed from the shop when completed?
5. What is the crane capacity? Is it sufficient to lift the finished product?
6. What is the shop capacity? If the order is likely to take up more than 50 percent of the shop floor, scheduling problems could arise.
7. Is the shop divided into bays for requirements such as isolating stainless steel from carbon steel?
8. Is the shop clean? A dirty shop could result in contaminated welds.
9. Is the shop cluttered? If so, dangerous working conditions and poor delivery are possible.
10. Is the shop laid out so that work follows a logical progression? If not, delivery delays may be encountered because of inefficiency.
11. Are blasting and painting facilities isolated from welding areas to prevent contamination?
12. Is there a suitable layout area for drawing and manufacturing sequence operations?
13. Is there a quarantine area for incoming parts to be held for inspection, and is there a quarantine area for nonconforming material?

14. Notice if the products on the floor are similar to those to be ordered. If so, access their quality.

15. Check the inventory area for plate, heads, flanges, nozzles, etc. Are they stored properly, and are they identified by stamping or some other permanent method? A small inventory may indicate a shop that is experiencing financial difficulties.

NOTE: Some shops may use what is termed a "Just-In-Time" (JIT) philosophy, wherein they receive components just in time for fabrication. Consequently, they carry little or no inventory. Many chemical, petrochemical, and refining companies prefer to purchase from fabricators that carry an inventory. This allows for inspection scheduling and a rapid response to emergencies.

16. Check the scrap bins for signs of excessive rejected material and parts.

17. Is fabrication taking place outdoors without proper protection? If so, consider that wind can blow away inert gas shields, and welding in the rain could lead to hydrogen cracking.

18. Does the shop have appropriate test facilities, including a clean water supply?

Personnel and Management Policy

Personnel and management policy are important factors that should not be overlooked. The following outline some general areas one should consider.

1. Are the workers busy or is there apparent slack time? Do they appear to be highly motivated or are they sullen?

2. Did senior management visit the shop floor during the tour? If so, how were they received by shop-floor personnel?

3. What quality and safety slogans or posters are in evidence? Do they reflect current philosophies?

4. Does the supplier have a properly functioning quality control or inspection department? Spend some time observing inspections on products that are in the shop at the time of the visit. Ask to examine some inspection and test records of past work. Are inspectors forceful and respected, or are they treated as a nuisance?

5. If inspection services are contracted, visit the inspection company's facilities to review its policies, procedures, and personnel qualifications.

6. Does the supplier have its own engineering staff and are they familiar with the applicable codes and standards for pressure equipment? Ask to see the codes and standards used for design. If they are out of date, problems could arise.

7. Does the shop have its own fabrication quality manual based on ASME B&PV Code, Section VIII, Division I, approved by the appropriate authorities? Does it have an applicable fabrication stamp?

8. Is the shop approved to fabricate equipment under other jurisdictional or technical associations such as API?[2]

9. Is the fabricator's shop unionized? If so, when does the contract expire? Does the supplier have a history of labor disputes?

10. Determine how equipment is to be shipped to the job site. Can trade union contracts for the shippers disrupt delivery? If so, alternative methods of shipment should be explored with the fabricator.

11. Obtain a list of subcontractors that the fabricator normally uses. Have any of them been evaluated as your suppliers? If not, evaluation visits to some of the major suppliers may be necessary.

---

[2] American Petroleum Institute, Washington, DC.

12. Review some of the subcontractor's procedures that have been submitted to the fabricator for previous work. If no procedures exist or were not submitted, problems may be encountered.

Welding

The following points outline welding considerations one should take into account when selecting a fabricator.

1. Examine the fabricator's welding procedures, making certain that they are registered and/or approved by the proper authorities. Determine whether or not the welding procedures are available in the shop to the welders who are performing the work; preferably at their work stations. Have the procedures been approved by your engineers?

2. Witness some of the welding in progress to determine its quality.

3. Review radiographs from previous work.

4. Examine the method of welding consumable control. Are low hydrogen electrodes maintained in a heater? Are there ovens for drying electrodes and are they being used? Are electrodes left lying around? If so, cracking problems will likely occur.

5. Determine what the longest weld that can be made with automatic equipment is and then decide whether it is compatible with the anticipated need.

6. Examine the fit-up on work in the shop. If excessive hammering is used to make the fit-ups, cracking and other welding defects could occur.

7. How is slag removed from weld deposits? If it is removed by grinding, undercut and other problems are possible.

8. Is welding properly shielded from contamination? Is it protected from drafts when doors are opened? Is it protected from paint, blast-cleaning compounds, and other airborne particles?

9. Are heat-treating facilities available on site or will such work need to be subcontracted?

Testing, Gauging, and Other Considerations

Testing, gauging, and other considerations one might wish to consider are as follows:

1. Are painting facilities located on site or is such work subcontracted?

2. Review all test procedures such as ultrasonics, radiography, hardness, impact, and hydrostatics to ensure that they are satisfactory for the completion of the work.

3. Examine gauge test and storage facilities to make certain that gauges are calibrated and maintained in good condition.

4. Ask to have a final tour of the shop starting at materials receiving. Follow the sequence of the fabrication through the facility to identify any potential problem areas.

5. Perform an assessment of inspection personnel after witnessing their performance during the visit.

6. Prepare a comprehensive evaluation report either recommending or rejecting the fabricator.

Check sheets facilitate the evaluation of suppliers as well as promote consistency if all inspectors use the same criteria. The following check lists and sheets have been used successfully in North America and Europe, and can be readily adapted to specific situations by the deletion of irrelevant items.

SUPPLIER EVALUATION CHECK LIST

SUPPLIER:
ADDRESS:

PLANT LOCATION:
INDUSTRY:
DATE:

	SHOP	OFFICE	OTHER
PLANT SIZE			
NO. OF EMPLOYEES			
UNION			
CONTRACT EXP. DATE			
NO. OF INSPECTORS			
QUALITY PROGRAM			
SAFETY PROGRAM			
CLEANLINESS			
COMPUTER CONTROL			
LIGHTING			

TEST EQUIPMENT

TYPE	YES/NO	COND.**	COMMENTS
HYDROSTATIC			
LEAK			
RADIOGRAPHIC			
MAG. PARTICLE			
LIQ. PENETRATION			
ULTRASONIC			
EDDY CURRENT			
ACOUSTIC EMISSION			
TENSILE			
IMPACT			
CHARPY OR IZOD			
HARDNESS			

EQUIPMENT

TYPE	YES / NO	SIZE / CAP.* / NO.	COND.	COMMENT
ROLLS				
SHEARS				
PUNCH				
LATHE				
DRILL				
SAWS				
PLANE				
SHAPER				
MILLER				
PRESSBREAK				
CRANE				
AUTO. CUTTING				
WELDING				
• SMA				
• MIG				
• TIG				
• SAW				
• EB				
• SPOT				
AUTO. AND / OR SEMI-AUTO. WELD				

* CAPACITY; ** CONDITION

SUPPLIER EVALUATION CHECK SHEET

TEST EQUIPMENT

	YES / NO	COND.	COMMENTS
PROFILE			
HOLIDAY			
AUTO. DIMENSION			
BORESCOPE / FIBERSCOPE			
MIRRORS			
GAUGES			
• PRESSURE			
• TUBE			
• THREAD			
• WELD PROFILE			
• WELD REINFORCEMENT			
• DEPTH			
MICROMETERS			
CALIPERS			
LEVELS			
SQUARES			
WET & DRY FILM			
METALLOGRAPHIC			
• POLISHING EQUIP.			
• PHOTOGRAPHIC			

EQUIPMENT

TYPE	YES / NO	SIZE / CAP. / NO.	COND.	COMMENT
JIGS				
WELD ROLLS				
ELECTRODE STORAGE				
OVENS				
WELDING SHIELDS				
BLAST CLEANING				
PAINT				

PLANT PRACTICE COMMENTS	
EQUIPMENT ON FLOOR:	WELDING PROCEDURES:
STORAGE YARD:	TEST PROCEDURES:
SCRAP BINS:	INSPECTION QUALITY:
WORK SCREENED FROM CONTAMINANT:	MANAGEMENT ATITUDE:
DOOR SIZE:	ENGINEERING DESIGN:
DOLLAR VALUE OF CONTRACTS/YR:	QUALITY MANUAL:
SHIPPING METHODS & MAJOR ROUTES:	U STAMP:
POWER SUPPLY INTEGRITY:	API QUALIFICATION:
WINTER SHIPPING RESTRICTIONS:	OTHER QUALIFICATIONS:
TEST WATER QUALITY:	
COMMENTS: SPECIAL INSTRUCTIONS:	

ADDITIONAL CONSIDERATIONS

	LOCATION	SIZE LIMITS	COMMENTS
HEAT TREATMENT FACILITIES			
HEAD MANUFACTURING			
FORGING			
CASTING			
DRAWING			
STAMPING			
EXTRUSION			

CONTRACT INSP. :

MAJOR SUB-CONTRACTORS:

INVENTORY OF WORK:

REFERENCES & DESIGN CODES:

INDUSTRY REPUTATION:

HEIGHT OR WIDTH RESTRICTIONS ON SHIPPING:

FINAL ASSESSMENT

1. Does plant have capacity to handle work?
2. Is equipment sufficient to complete job?
3. Is quality of work in shop adequate?
4. Are inspectors capable and well equiped?
5. Is quality program adequate?
6. Is testing equipment adequate?
7. Is welding equipment and quality adequate?
8. Is sub-contractor control adequate?
9. Are additional facilities such as heat treatment available and adequate?
10. Is management committed to quality

NOTE: A "no" response to any of the above questions would disqualify the supplier.

INSPECTOR: DATE:

CHAPTER 5

General Good Practice

Overview

Inspection to ensure the quality of pressure vessels and piping is aided if certain general rules or suggestions termed "good practice" are adopted as applicable by designers, fabricators, and inspection personnel. Good practice is outlined under the following four headings. Many of the points outlined are Code suggestions, while others are items that have been established through experience.

Design

The following good practice principles apply to design.

1. Minimum thickness of 1/4 in. (6 mm) for shells and heads, except for those made from certain specialty alloys, which may be thinner.
2. Welds are generally separated by 1 in. (25 mm) of parent metal; i.e., do not place weld upon weld.
3. Longitudinal seams are located clear of nozzles, lugs, manways, reinforcing pads, support welds, and other obstructions.
4. All longitudinal and circumferential weld seams are located so that they may be inspected with vessel internals in place (except for heat exchangers and other similar equipment).
5. Longitudinal welds are staggered 180°, if possible, on large vessels.
6. Circumferential welds are located clear of all nozzles, manways, reinforcing pads, support welds, etc. If a reinforcing pad must cover a weld, the weld is normally ground flush and radiographed or otherwise tested before it is covered.
7. The minimum size of internal fillet welds functioning as strength welds is 1/4 in. (6 mm).
8. Bolt holes straddle center lines.
9. Corrosion plates (or pads) wider than the saddles are used when horizontal vessels are to be supported directly on concrete pads.
10. Taper grinding at a 1-to-4 ratio is recommended on the inside of vessels and at nozzles where differing thicknesses occur. However, end grain attack, encountered in some industries, may require taper grinding to be performed on the outside of vessels. Alternatively, the exposed edge may be buttered with weld metal.
11. The internal edges of nozzles are generally rounded to a 1/8 in. (3 mm) minimum radius, except for drain nozzles which are made flush.
12. No welding is permitted after post-weld heat treatment, except for seal welding of threaded openings, which may be performed after heat treatment.
13. Location of all weld seams, nozzles, openings, internals, etc., are indicated on shop drawings by the distance to a common reference line.

14. Davits are provided for manhole covers and are mounted so that covers can be swung completely away from the opening.

15. Seamless pipe is used most often, but not always, for nozzles.

16. On lined vessels, the lining is carried out over the flange faces of all nozzles and other openings.

17. Minimum wall requirements, rather than average wall thickness, are often preferred for exchanger tubing. Average wall is acceptable, provided the design is adequate.

18. Tubesheet thicknesses are generally minimum requirements.

Testing

Good practice procedures for testing are as follows:

1. Removable internals are installed after stress relieving.

2. Hydrostatic test pressure is maintained for a 30 min minimum.

3. Vessels are not painted or insulated prior to hydrostatic testing.

4. Hydrostatic test water is heated above 60°F (16°C) to prevent possible cracking of carbon steel components.

5. Hydrostatic test water contains low chlorides, sulfides, and other contaminating compounds. Demineralized water may be used to test austenitic stainless steels in order to prevent chloride ion stress corrosion cracking (SCC).

6. After the final hydrostatic testing, vessels and piping are dried immediately and thoroughly cleaned—both internally and externally. This helps to control microbiologically influenced corrosion (MIC). Water is not permitted to stand in vessels for more than 8 to 12 hours.

Identification

The following is a list of good practice guidelines designed to aid one in the identification process.

1. Plate or material identification includes:

   a) material, grade, and classification, e.g., ASME SA 516 grade 70 or SA 240 type 304 or API 5L;

   b) heat and slab numbers;

   c) mill identification;

   d) size for pipe and fittings; and

   e) indication of manufacturing process for pipe and fittings; e.g., from the API requirements—seamless (Type S) or electric resistance welded (Type E).

2. The nameplate is located above the manway and outside any insulation or other covering.

3. The nameplate is manufactured from austenitic stainless steel or some other corrosion-resistant alloy.

4. The nameplate, as a minimum, contains:

   a) name of the manufacturer;

   b) date of manufacture;

   c) vessel registration number;

   d) manufacturer's serial number;

   e) design temperature;

   f) design or maximum operating pressure;

   g) material of construction;

   h) code section governing fabrication; and

   i) manufacturer's authorized code stamp.

5. There is some method of identifying the valve position such as a rising stem, direction arrow, or some other suitable means.

Protection and Storage

The following outline good practice procedures for protection and storage.

1. All flanged openings without covers are protected with plates, flange protectors, or other suitable materials.
2. Uncoated steel parts are protected with suitable oils or greases during shipping and storing.
3. Components stored outdoors prior to shipping are sloped to promote water drainage.
4. Materials stored for six months or more before shipment are placed on gravel or paved pads. Wooden bolsters, used for storage, are treated to prevent degradation and moisture accumulation. Appropriate air space is maintained between adjacent components to prevent moisture accumulation. Materials are stored away from obvious sources of contamination such as road salts, salt-water splash zones, industrial stack plumes, etc.

CHAPTER 6

Shop Inspection of Pressure Vessels

Inspection of Pressure Vessels

Pressure vessels are fabricated in many shapes and sizes. They are constructed from a variety of materials, operate in varied process environments, and their fabrication can be complex depending on the alloys utilized. Some alloys or components may not be amenable to inspection techniques. For instance, magnetic-particle examination cannot be used on nonmagnetic material. In addition, some specialty alloys may require welding under vacuum conditions or other appropriate shielding techniques.

Nevertheless, the inspection techniques employed to ensure pressure vessel compliance with purchase specifications are generally the same for all shapes, alloys, and pressure components used. For example, specifications for carbon steel will be the same as those of exotic alloys. Specifications for a reactor will be the same as those for heat exchanger shells and storage vessels. Consequently, the inspection of pressure vessels can be outlined in a basic guide with additions for specialty items. The approach outlined in this section has proven effective in shop inspection of pressure equipment fabricated from carbon steel, low-alloy steels, high-alloy steels, ferritic alloys, austenitic stainless steels, and some high-nickel alloys. However, to be effective, all preliminary activities (supplier evaluation, pre-inspection meeting, etc.) must have been performed.

Shop inspection of pressure vessels and piping by the purchaser's representative during fabrication is the best way of ensuring that components meet the purchase specification requirements. The purchaser's representative, normally called an inspector or a quality surveyor depending on the type of program in force, will usually visit the fabricating shop several times to:

- witness inspections by fabricating personnel;
- examine the results of any nondestructive testing (NDT) performed;
- review documentation associated with the contract; and
- ensure that proper packaging, loading, and shipping instructions have been followed.

In some instances, the purchaser's inspector may find it necessary to perform certain tests personally in order to confirm critical measurements, verify NDT results, and/or ensure that all jurisdictional and code requirements have been met.

NOTE: See "Pressure Vessels: The ASME Code Simplified," in the Bibliography (Appendix 8).

Inspection Equipment. Inspectors may utilize either their own testing instruments or those of the fabricator if satisfactory equipment is available. An inspector's kit can include the following:

1. flashlight (explosion-proof if required);
2. scraper (1 in. [2.54 cm] minimum width);
3. inside calipers, both US Customary (English) and Metric units;

4. outside calipers, both US Customary (English) and Metric units;
5. steel rule (minimum 20 ft [6.1 m] length);
6. telescoping tube gauges;
7. hand lens (5 to 10X);
8. steel straightedge (minimum 4 ft [1.22 m] long);
9. steel square (4 ft x 2 ft [1.2 m x 6.1 m] minimum);
10. level (4 ft [1.22 m] minimum) or string level;
11. yellow marking crayon;
12. weld profile gauge;
13. depth gauge;
14. micrometer;
15. gloves;
16. coveralls;
17. safety footwear;
18. safety glasses;
19. notebook and pen;
20. cameras (35 mm and Polaroid®);
21. compass or protractor;
22. wet and dry film gauges; and
23. small file.

<u>Optional Equipment</u>. Other helpful supplies include:

- coil of piano wire;
- vernier or dial calipers;
- theodolite (survey equipment);
- various nondestructive evaluation (NDE) equipment (thickness gauge, flaw detector, magnetic particle test yoke, liquid penetrant, etc.);
- set of welding radiograph standards;
- profilometer;
- portable hardness tester; and
- blast cleaning or other surface preparation standards;

Supplies such as hydrostatic test gauges, radiograph viewer, etc., are customarily supplied by the fabricator.

Inspector Preparation. After the preliminary work has been completed (supplier evaluation, bid evaluation, finalization of contract documents, and contract issuance) the inspector prepares for the next phase by opening a job file in which the following information can be found:

1. purchase order and all amendments;
2. approved drawings;
3. job specifications;
4. quality program requirements;
5. fabricator's welding procedures to be used on the job;
6. verify NDT procedures that are to be used by the fabricator;
7. jurisdictional requirements;
8. proposed fabrication schedule; and
9. name, telephone, and (FAX) number of contact personnel at the fabricator's facility.

Pre-Inspection Meeting. A pre-inspection meeting is usually arranged after the contract is let and a purchase order has been issued. This meeting between the customer and fabricator inspectors is used to verify that all specifications and quality requirements are understood. The following steps are important aspects of this process.

1. All purchase order and specification requirements are reviewed, frequently clause by clause.
2. All code requirements are reviewed.
3. Guidelines are established for witnessing or for waiving special requirements.
4. Fabricator's subcontractors are identified and quality and/or inspection requirements established.
5. Welder certifications are established and confirmed.
6. Special qualification or welder demonstration tests are requested.
7. NDE qualifications and requirements are established and confirmed.
8. Fabricator's test facilities are reviewed.
9. Office space at the fabricator's facility is arranged for the purchaser's inspector.
10. Inspection and test plans are submitted by the fabricator. Witness and hold points are established.
11. Requirements for a trial assembly, as applicable, are established.
12. Document requirements (e.g., mill test certificates) are outlined.
13. Acceptance/rejection criteria are specified.
14. Preservation, shipping, and storage requirements are outlined.
15. Methods of resolving nonconformances are established.

Inspection Visits. The number and extent of inspection visits vary with the complexity, cost, and importance of the equipment. Obviously, an inexpensive, easily replaced vessel for a noncritical service will require less extensive inspection than a vessel that could either affect an entire unit or result in hazardous conditions in the event of failure. Regardless, inspections are planned to coincide with events that will provide the most useful information. A minimum of three visits are needed and should include the following:

1. examination of major raw material;
2. fit-up of major components and the commencement of welding;
3. observation of some NDT;
4. random verification of major dimensional elements;
5. ongoing visual inspection of welding and other fabrication elements;
6. witnessing of hydrostatic testing;
7. witnessing of trial assembly, if applicable;
8. witnessing of functional testing, if applicable;
9. verification of surface finish, as applicable, including preparation for painting and examination of coatings;
10. review of all documentation such as mill test certificates, welder certification, NDE reports, radiographs, etc. Confirmation that affidavit of manufacture has been completed and assurance that an accurate nameplate has been provided;
11. final inspection to verify that all loading and shipping requirements are met; documents are complete and acceptable; fabrication has been accepted by state, provincial, or other jurisdictional authorities; and the equipment is ready to be shipped; and
12. inspection for damage when equipment is received at the site.

If problems are encountered during fabrication, inspection visits are increased. If the inspector needs to become a virtual resident, consideration may be given to removing the equipment for completion by another fabricator.

Inspection of Raw Materials

Guidelines for the inspection of raw materials are as follows. The first inspection visit is usually timed to establish that the major raw materials have been received and are acceptable.

1. Review the fabricator's drawings, comparing them with those supplied to the purchaser.
2. Using the bill of materials from the drawings and the specifications outlining what materials are to be used, verify that the correct quantity and type of product has been ordered.
3. Confirm plate thickness and examine the shell plates for pits, gouges, cracks, heavy scale, or other signs of damage. Request that any heavy scale be removed so that the underlying metal can be examined. Check for contamination of high-alloy material, if applicable.
4. Record the plate identification, ensuring that it is stamped or stenciled on each item. If plates are cut, check that identification has been transferred to each piece of plate.
5. Check to be certain that mill test certificates comply with the specifications. If mill test certificates are not available until a later inspection date, the plate may be released for fabrication with the understanding, confirmed in writing by the fabricator, that it will be removed from the process if the mill test certificates are not received or prove unsatisfactory. Witness any tensile, impact, and ultrasonic flaw detection tests (at material supplier, as necessary).
6. Examine heads to ensure that they are the correct type. Measure the head diameter and verify the thickness. Check knuckle radius, straight flange length, and other specified measurements. Verify that the identification is stamped or stenciled on each head.

NOTE: See the detailed section on head inspection.

7. Ensure that plate and flanges are stored properly. The plate should preferably be stored in racks separated by spacers to prevent water accumulation and crevice corrosion. The heads should be stored on platforms or racks so that water can not accumulate. Plate stored flat with no spacers can result in extensive pitting in carbon steels and some austenitic alloys, particularly if deicing salts can splash on them or the shop is located in an industrial environment.
8. Measure flange thicknesses and check dimensions to ensure that they are the flanges specified. Verify that all flanges are properly identified.
9. Check the flange finish, e.g., raised face, ring, and groove.
10. Check flange bolt holes to ensure they are the correct size, have been properly drilled and tapped, have the correct number of bolt holes for the flange size, and are properly spaced and located around the flange (bolt holes generally straddle the center line).
11. Confirm the diameter and thickness of any nozzle piping.
12. Check to make sure that all pieces of nozzle piping bear the correct identification. Check that nozzle piping is seamless, if so specified.
13. Check manhole components if they are different from nozzles (for marine manways, etc.). Determine that dimensions are accurate and that there is no taper. Check cover plates to ensure proper gasket seating surfaces.
14. Examine bolting. Verify inside diameter (ID) including manufacturer's mark and bolt grade,

which should be stamped on bolt ends. Ensure that bolts are the correct size and length.

15. Check that nuts are the appropriate grade for the bolting. Ensure that nuts will thread on bolts by sampling a number.
16. Check gasketing material, as applicable.
17. Heat Exchangers. The following factors should be accounted for regarding heat exchangers.
    a) In addition to the shell plates, check the channel, baffle, and cover plates, as applicable. Ensure that each plate contains the proper identification.
    b) Check exchanger tubes for length and number. Problems often arise when a few tubes of different material are added because the order was a few tubes short.
    c) Check tube ID and outside diameter (OD).
    d) If tubesheets and baffle plates have been drilled at this point, check the holes for size, grooving, finish, etc. Ensure that baffles are stacked so that the holes line up.

Additional details on the inspection of raw materials for certain components are provided in the following sections.

Inspection of Heads. Formed heads require considerable inspection during the fabrication of pressure equipment. They may be formed by a number of processes such as spinning, pressing, forging, and/or shaping. Each forming operation can introduce defects.

Conventional heads are of three basic types: hemispherical, ellipsoidal, and torispherical (commonly referred to as the ASME Flanged and Dished Head). There are other less frequently used types: conical (no knuckle), toriconical (with a knuckle), flat heads, and heads bearing slight curvatures so that they are not flat, yet they do not reach the dimensions of a fully dished head. Consequently, it is important that the inspector determines the type of head intended and its dimensions.

Ellipsoidal heads are commonly used for small and medium diameter vessels, while large vessels utilize hemispherical or torispherical heads. However, practice varies from plant to plant. Some companies use torispherical heads for small vessels, while others do not use this type of head at all. The heads may be seamless or of welded construction. Two plates may be joined to make a larger section from which to form the head or the design may require a complicated, segmented construction in which a central circular plate, "dollar plate," is joined to a series of alternating triangular shapes to form an “orange peel” construction. In all cases involving welding, weld inspection, with the same standards and considerations as those needed for the shell construction, is an obvious requirement.

In general, all heads except hemisphericals are used with straight flanges. Formed heads that are not thicker than the shell and are butt-welded to the pressure vessel shell do not require straight flanges. The usual length of straight flanges is 2 in. (50 mm) for ellipsoidal heads, 1.5 in. (37.5 mm) for torispherical heads, and 0.0 in. (0.0 mm) for hemispherical heads.

Straight flange lengths for heads thicker than the shell are generally the greater of 2 in. (50 mm) or the wall thickness for ellipsoidal heads, the greater of 1.5 in. (37.5 mm) or the wall thickness for torispherical heads, and 0.0 in. (0.0 mm) for hemispherical heads.

Inspection of heads usually entails the confirmation of dimensions, including fabrication tolerances, verification of materials, and testing as follows:

1. Make certain that the proper identification is present on the head or each segment of the head, as applicable.

2. Examine the mill test reports to ensure that they comply with the specification and that they can be traced through identification to the plates used in manufacturing.

3. Check that the appropriate head has been fabricated, e.g., ellipsoidal, hemispherical, torispherical, etc.

4. Examine the inside and outside surfaces for damage, pitting, heavy scale, overheating, cracking, etc. Require that heavy scale be removed by grinding, sandblasting, or other suitable means to check the underlying surface.

5. Check the deformed areas for signs of defects such as cracking, gouging, burning, etc.

6. With the head on a flat plate, open side up, check the inside depth of the dish (d) by placing a straightedge across the head and measuring the greatest depth with a solid steel rule. A plumb bob suspended from the straightedge may be used to ensure that a true vertical height is used. Tolerances are usually as follows:

   - 0.5 in. (12.7 mm) to 0.0 in. (0.0 mm) for heads up to 48 in. (122 cm) in diameter.
   - 0.75 in. (19.0 mm) for heads over 48 in. (122 cm) and up to 96 in. (244 cm) in diameter.
   - +1 in. (25 mm) to 0.0 in. (0.0 mm) for heads over 96 in. (244 cm) in diameter.

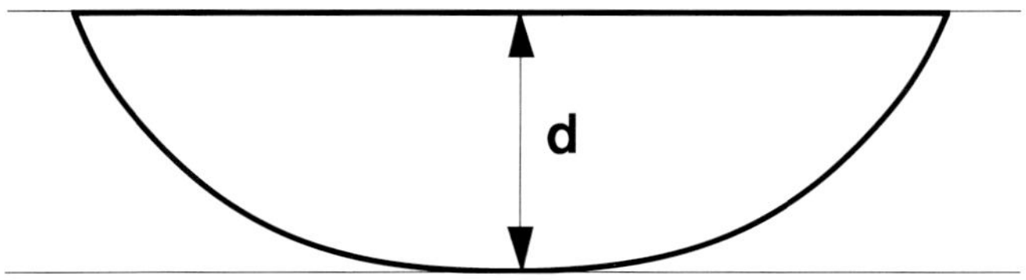

7. Measure the diameter of the head. The ID is measured on hemispherical and ellipsoidal heads, while the OD is measured on torispherical heads. The diameter is determined by placing a steel rule across the head and moving one end of the rule through a small arc, while holding the other fixed to record the maximum length. Readings are taken in four locations: at an arbitrarily established zero point, then at 45°, 90°, and 135° from that point around the circumference. D max–D min is usually limited to less than 1.25 percent of the diameter of the matching vessel shell or skirt, as applicable.

8. As necessary, measure the outside circumference of the head to determine whether or not it will match the outside or inside shell or skirt circumference.

9. Check whether a flange is required. The length of any flange can be determined by placing a steel square on a flat plate or other suitable surface, open side down. Only that part parallel to the vertical leg of the square is identified as the flange.

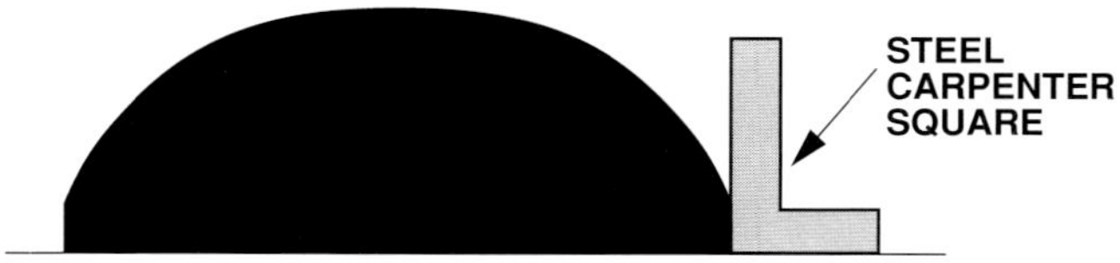

10. On torispherical heads, determine the actual crown radius by placing the upturned head on a flat plate and inserting a straightedge of equivalent length to the maximum chord for the size of head specified. The crown radius is then:

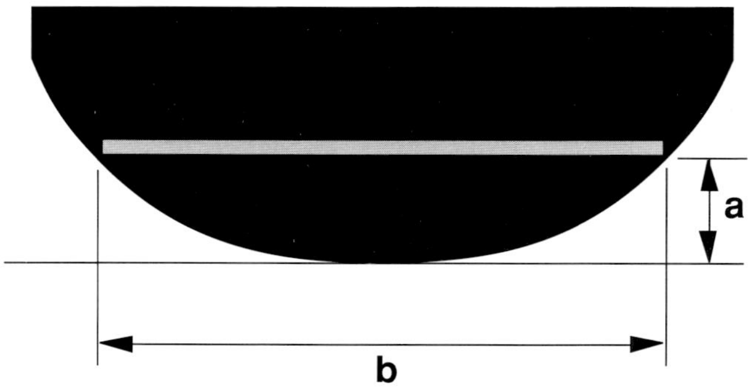

$$h = \frac{a}{2} + \frac{b^2}{8a} \text{ inches}$$

where **a** = height of the straight edge above the crown of the head, and
b = length.

11. Check the knuckle radius of the heads using templates provided by the manufacturer.

12. Measure the thickness of the heads using dial or vernier gauges, micrometers, calipers, or some other reasonably accurate instrument. Measuring with a steel rule is unacceptable. Measure the thickness in each quadrant around the circumference.

NOTE: Use ultrasonic thickness measurements to determine the minimum thickness of the flange and knuckle areas. In addition, the minimum thickness of the crown should be established as it often suffers considerable thinning during forming, particularly if a nozzle has been installed. The minimum machined thickness permitted is generally 90 percent of that of a blank head.

13. For ellipsoidal heads, check that the standard ratio, usually 2-to-1, has been maintained. This can be done by placing a straightedge across the open end of the head and measuring the head diameter (D), height (h), and the flange length (l). The head ratio is:

$$D : 2(h-l)$$

where: **D** = head diameter.

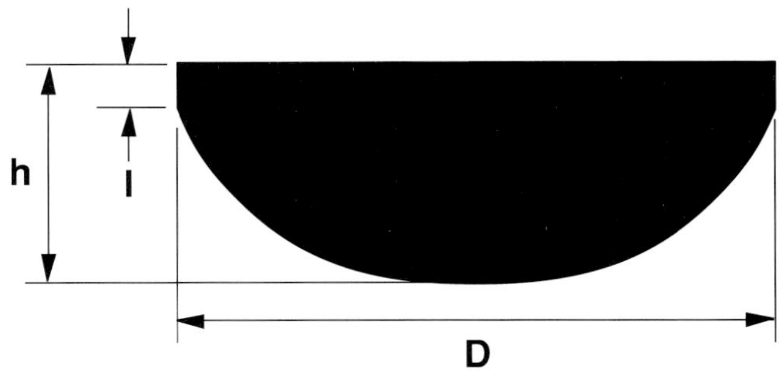

14. Check that the heads have been heat treated as required by examining the test reports and heat treat charts. Cold formed heads are generally annealed, but hot formed heads are generally stress relieved. Exceptions involve some high-alloy materials.

15. Inspect any welding in the same manner as for vessel fabrication. If spot radiography of segmented heads is specified, examine a spot on each weld joint, preferably in locations of high stress such as at T-junctions or where the greatest deformation has occurred. Obvious surface imperfections such as undercut, porosity, or blowholes require correction.

Inspection of Flanges and Fittings. Flanges and fittings are generally not considered critical components in the fabrication of small pressure vessels. Nonetheless, they can affect the integrity of the pressure envelope leading to premature failure, costly replacement, production loss, and higher maintenance costs. Therefore, flanges and fittings are usually subjected to a visual examination with dimensional verification on a sampling basis. If defects are found, the degree of inspection is normally increased.

Flanges and fittings are supplied in carbon steel and other alloys. A list of inspection procedures is provided below. Inspection requirements, except for certain NDT applications, are similar regardless of the alloy.

1. Components are examined for scale, laps, laminations, gouges, pits, and other surface imperfections. Examination with hand lenses is particularly effective for identifying defects.

2. Confirm the material and its identification through examination of mill test certificates, application of chemical spot tests, determination of ferromagnetism, etc.

3. Examine the components to ensure that they bear the correct identification.

4. Confirm with dimensional checks such details as flange thickness, hub thickness, concentricity, gasket facing, bore, etc.

5. Examine bolt holes to ensure that they straddle center lines and are the right size.

6. Examine surfaces to ensure that there are no machining defects such as tool chatter or tears.

7. Gauge ring-type joints to ensure correct ring dimensions.

8. Check threaded surfaces with a go/no-go thread gauge. Inspect threads for tears, cracks, flat spots, etc.

9. Confirm that any required heat treatments have been performed.

10. Check hardness and witness charpy tests, ultrasonic tests, and other such specification requirements.

11. Review radiographs, if applicable.

Inspection of Valves. Valves are essential components in pressure systems. Although there are many types of valves (gate, globe, butterfly, ball, etc.), the following basic steps are common to all. Additional requirements may apply depending upon the valve specified. Gasketing, packing, and other such components may require examination as well.

1. Verify that the type and size of valves obtained are those outlined in the specification.

2. Check the valve dimensions to ensure that they meet specification requirements and confirm the following:

    a) face-to-face dimensions;

    b) inlet and outlet bore dimensions;

    c) flange thicknesses and diameters;

    d) bolt hole size and number;

    e) body and bonnet thickness, as applicable;

    f) valve stem diameter, as applicable;

    g) gasket facing, ring grooves, etc., if applicable; and

    h) weld preparation for butt weld ends.

3. Visually inspect the surface finish of valve bodies, i.e., bonnets and other components, to ensure that there are no casting defects such as scabs, cracks, excessive scale, pits, holes, or shrinkage.

4. Examine valve seating surfaces to verify proper surface finish, i.e., free of pits, gouges, chatter marks, etc.

5. Inspect mill test certificates to confirm compliance of all component material (e.g., for body, bonnet, trim, gate, disc, and ball elements).

6. Examine bolting mill test certificates.

7. Look at bolting to ensure that all bolts have been properly identified. Witness the required NDE such as magnetic-particle or liquid penetrant on a representative sample of bolting.

8. Verify that the valve type, size, material, and manufacturer data is evident on castings or on tags permanently attached to the valves.

9. Examine radiographs of valve bodies, as applicable, to ensure compliance with specifications. Care must be exercised in the interpretation of film because some valve standards allow for casting defects that are virtually through the valve wall. In addition, for criterion purposes, a fitness may be required. Rejecting a valve because of excessive porosity in an external reinforcing rib that has no effect on the pressure retaining aspects, makes little economic or practical sense.

10. Witness hydrostatic tests on valves, including

seat leak tests, opening and closing valves against pressure, etc.

11. Witness tests on any actuators to establish:

    a) opening and closing times;

    b) power, pressure, etc., required to open and close; and

    c) open and closed positions;

12. Following actuator tests, inspect valve components for signs of excessive wear or jamming.

13. For valves requiring repair, witness NDT on excavated areas to ensure that the defect has been removed.

14. Verify welder certification.

15. Examine testing of repairs including a review of radiographs, as applicable.

16. Ensure that any required heat treatment has been performed after repairs.

17. Ensure that valve openings on large valves are covered with suitable hard covers to prevent damage during shipping. Small valves may be packed in boxes or crates. Shipment of loose valves is not recommended.

Bolting. Although bolting is often overlooked during the inspection of new pressure equipment and piping, bolting failure can result in production loss and hazardous conditions. Bolting is generally inspected on a sampling basis, usually 10 to 20 percent of a lot, but there has been a tendency to increase the sample size because of nonspecification material failures. Now, 100-percent inspection is often used for critical items where leakage or failure could result in serious economic losses, hazardous conditions, or the contravention of pollution and other environmental regulations. Even then, some defects may slip through. Consequently, it may be necessary to inspect critical items at least twice in an attempt to catch all defects.

(3)American Society for Materials and Testing, Philadelphia, PA.

Bolting inspection generally entails the following:

1. Examination of the bolt and nut identification. Identification is stamped on bolt heads or on the ends of studs and nuts. It includes the manufacturer's identification, the bolt or nut specification, type and grade; e.g., ASTM(3) A 193 B7 and A 194 2H for bolts and nuts, respectively. Some purchasers insist that studs be stamped on both ends for ease of identification. It is advisable to avoid bolts and nuts that are not identified properly and those of suppliers whose symbols cannot be identified.

NOTE: When the ASME B&PV Code accepts an ASTM standard, the specification is designed with an "S" prefix, e.g., SA 193 B7.

2. Inspect nuts and bolts for dirt, rust, scale, thread damage, and other signs of degradation.

3. Measure the length and diameter of these items to ensure that they meet the specification. Measure the bolt pitch diameter.

4. Using a thread gauge, confirm that the appropriate thread has been produced on bolts and nuts.

5. Verify that the nut thickness is adequate.

6. Examine mill test certificates to confirm chemical and physical properties.

7. Examine bolts and nuts, particularly those that have been heat treated, for signs of cracking, burning, distortion, etc.

8. Some purchasers require a metallographic analysis to ensure compliance with heat treatment. Examine photomicrographs, comparing them with relevant standards.

9. Witness magnetic-particle, liquid penetrant, hardness, or other NDT, if applicable.
10. Witness nut stripping tests, if necessary.
11. Ensure that nuts and bolts requiring long storage periods are suitably protected to prevent corrosion.
12. Witness the threading of some nuts on bolts, then examine threads for galling or other signs of damage.

NOTE: See Appendix 2 for reference information.

Inspection of Tube-and-Shell Heat Exchangers

Heat exchangers also come in a variety of shapes and sizes. They may have fixed tubesheets so that the bundles cannot be removed or they may have removable tube bundles—often these are of floating head construction. The tube bundles may be single pass (in which the flow goes in one end and out the other) or they may be of multiple pass construction, using bent tubes and/or baffling to change flow direction. Therefore, it is important for the inspector to ensure that a complete information package, including drawings, has been obtained before trying to ascertain what type of inspection is required. For example, the examination of tube bending would not be applicable for single-pass exchangers.

Applicable inspection guidelines for heat exchangers are outlined below.

General. The following is a list of general heat exchanger specifications one should look for.

1. Examine the drawings and specifications to determine what type of inspection is required.
2. Examine the raw materials (refer to the "Raw Materials" section).
3. Examine the heads (see "Inspection of Flanges and Fittings").
4. Examine valves, fittings, and flanges (see both "Inspection of Flanges and Fittings" and "Inspection of Valves").

Exchanger Shells. When checking exchanger shells, verify the following:

1. Examine the rolled shell sections, particularly the plate edges, for cracks, tears, breaks, and other defects.
2. Ensure that weld preparations have not been damaged during rolling.
3. Check the weld edge alignment prior to welding and look for signs of hammering to round up the shell.
4. Check plate thickness after rolling.
5. On fixed tubesheets, check the baffle locations, hole diameter, and alignment of holes before closure welding.
6. Check that the appropriate welding procedures and techniques are used.
7. Examine welding (see previous section on welding).
8. After both the inside and outside surfaces are welded, check the shell for distortion or "out of roundness" by passing a template through it.
9. Check that the nozzles have been located as indicated on the drawing. Ensure that the holes cut for nozzles are not oversize.
10. Check that the nozzles are perpendicular to the shell and that they have been welded on both the outside and inside.
11. Confirm the overall length of the shell.
12. Place a level across flanges to check for misalignment. Use a square to check the flange

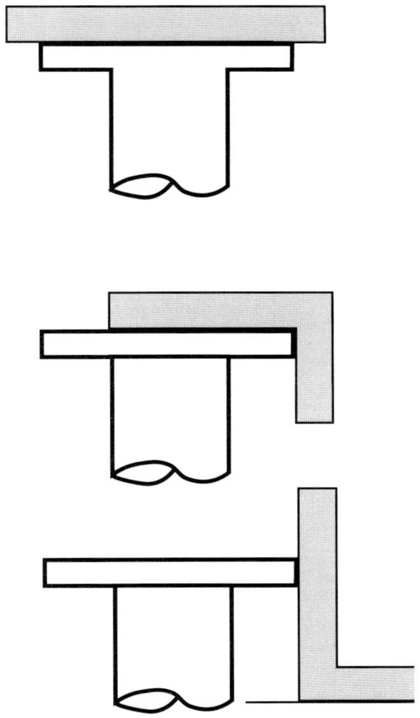

dimensions and alignment as well. (See figure above.)

13. Check flange finish and ascertain that bolt holes are drilled correctly, straddling center lines.

14. With multiple units, check that match marks, flange alignment, and all other locating indications are adequate and accurate.

15. Ensure that longitudinal welds are staggered, if applicable.

Tube Bundles. Tube bundles should adhere to the following specifications.

1. Check the number, length, OD, and ID of the tubes. Check some tube bending, if applicable, to detect signs of damage such as splitting, buckling, and/or wrinkling.

2. Verify the tube material, comparing it to mill test certificates.

3. Confirm the diameter and thickness of the tubesheets.

4. Confirm the diameter and thickness of the baffles.

5. Check the tubesheets and baffle holes for size, grooving, finish, and pitch, as applicable. Request that the baffles be stacked to make certain that the holes line up.

NOTE: Square pitch facilitates inspection and cleaning.

6. On clad or lined tubesheets, ensure that the cladding material is not contaminated and is of the thickness specified. Verify cladding material and witness tests to ensure proper bonding.

7. Witness the rolling of some tubes to ensure that proper expansion is obtained. Confirm the depth of rolling. Some specifications require 100 percent of the tubesheet thickness; others require 100 percent depth minus $^1/_{16}$ in. (1.59 mm) or $^1/_8$ in. (3.17 mm) from the back side of the tubesheet.

8. Examine the tube ends and the inside surfaces of the tubes for signs of cracks, tears, and overexpansion in the rolled areas. Check for signs of rolling or expanding beyond the back of the tubesheet.

9. Check the ligaments on the tubesheet for signs of cracking or other damage and check tubesheet flatness. Some form of crack detection may be necessary.

10. Witness the insertion of bundles into the shells.

11. Check the fit of the floating cover, channel, channel cover, and shell cover, as applicable.

12. Verify longitudinal baffling on multi-pass bundles, if applicable.

13. Check welding of tubes to tubesheet, if applicable. Ensure that the weld size meets specification requirements and that the welds are free

of surface defects. Check that welds have not penetrated the tube wall.

Final Inspection. Final inspection should include the following:

1. Check all flange faces to ensure that they are free of gouges, scratches, dirt, scale, etc.
2. Check the squareness of flanges.
3. Verify that tape or some other unacceptable aid is not used to hold gaskets in place during final assembly.
4. Check all couplings to ensure that they are free of dirt, scale, and other materials that may block them. Check couplings and other fittings for thread damage.
5. Witness the hydrostatic test on the tube bundles, verifying that the correct test pressure (generally 1.5 times design) is obtained. Ensure that pressure is held for a minimum of thirty minutes. Check tubesheets for bulging after the test. Ensure that the bundles are drained completely.
6. Oversee the hydrostatic test on the shell, then examine all welds. The pressure (again, generally 1.5 times design) should be held for a minimum of thirty minutes. The shell must be drained completely.
7. After the channel cover is attached, witness the final joint test.

NOTE: Ensure that bundles and shells are properly drained and dried to prevent microbiologically influenced corrosion and other problems.

8. Confirm the overall dimensions.
9. Using a protractor, compass, or some other suitable instrument, confirm the orientation of the nozzles.
10. Check the nameplate to ensure that it contains the required data and is not covered by insulation or other obstructions.
11. If a pneumatic or other leak test is performed, consideration can be given to filling the other side of the exchanger with water to prevent buckling. Exchanger joints may be tested by applying a soap solution, this will indicate leaks with bubbles. Caution must be exercised in the use of pneumatic tests, e.g., the equipment should not be approached until the test pressure has been reached and is holding steady.
12. Check that all bolts or studs bear the correct identification stamping and are of the correct size.
13. Check that all nuts are fully threaded onto the bolts.
14. Affirm that all nozzles requiring reinforcing have been reinforced.
15. Make certain that all reinforcing pads contain an adequate weep hole.
16. Confirm the location of saddle supports or legs if applicable.
17. Check the welding on the supports for cracks.
18. Check the surface finish, ensuring that appropriate cleaning has been completed before painting begins.
19. Make sure that the proper paint or other protective system is to be applied.
20. Observe the application of paint, insulation, etc., to ensure that it complies with the specification.
21. Make certain that any crating, packaging, or other preparation for shipment has been completed.
22. Observe the loading of the equipment for shipment.

23. Ensure that all required documentation such as drawings, test reports, mill test certificates, and affidavitis of manufacture are acceptable and complete.

24. Inspect units at the plant site for possible shipping damage such as bent nozzles, paint or insulation damage, road salts or other contamination, and dents or other indications that the unit has been dropped.

Inspection of Drums, Spheres, Reactors, Towers, etc.

Despite the varying size, complexity, and materials of construction, shop-fabricated pressure vessels are all inspected using a few basic techniques. These techniques are based on a visual confirmation that fabrication processes permit the construction of the component without causing damage by such factors as overheating, excessive cold working, and excessive restraint leading to scaling, cracking, distortion, etc. Assuming that all of the preliminary reviews, meetings, and other details have been completed, shop inspection entails the following:

- Verification that the major raw materials meet specification requirements.
- Confirmation that all welding, machining, and other fabrication operations are performed to written, approved procedures by individuals who possess the necessary skills and certification to complete the operation.
- Witnessing or performing tests to ensure that all components meet the required specifications including dimensional, physical, pressure retention, etc.
- Ensuring that all documentation pertaining to the fabrication and acceptance testing of the vessel has been completed and is acceptable.
- Ensuring that the vessel arrives at its destination in good condition.

Raw Material. The first inspection visit is usually timed to permit examination of major raw materials before fabrication starts. The following are involved in the inspection of raw materials.

1. Examination of plate, pipe, heads, flanges, valves, fittings, bolting, etc., to ensure compliance with specifications. Mill test certificates are examined and material identification is verified against these certificates.

2. The examination of plate, heads, and other material as outlined in this section, for signs of excessive scaling, pitting, overheating, distortion, etc.

3. Ensuring that all raw materials are identified by stamping, permanent stencilling, casting marks, etc., as applicable.

4. If materials are cut, make certain that identification is transferred to each piece. This should continue throughout fabrication at any time that material is being cut.

5. Measuring the material thickness. On plate, at least three spots along each edge are measured using calipers, vernier calipers, micrometers, or some other reasonably accurate measuring instrument (a ruler is unreliable). The thickness of plate, heads, etc., may be measured in the center regions using ultrasonics.

NOTE: The specified thickness may include a corrosion allowance. This corrosion allowance is compared against the specification. For austenitic stainless steel and other high alloys, a corrosion allowance is generally not required but there is one if manufacturing tolerances are on the positive side. Some companies use a minimum thickness criterion for such components so that the manufacturing tolerance provides extra thickness.

6. Measuring for thickness, diameter, bore, etc., of flanges.

7. Examination of a representative sample of fittings to ensure freedom from defects. Exami-

nation using magnification of 5 to 10X may be helpful. Ensure that the correct product, e.g., seamless rather than welded, has been obtained. Also, verify that the manufacturer is an approved supplier.

8. Verification that the right amount of each component has been received. Problems can result when items are shipped later to make up for a short order.

Inspection Prior to Welding. The following items need to be confirmed or approved before welding takes place.

1. Examine welder certification, confirming that it is current and reflects the techniques to be used on the job.
2. Examine the welding procedures to ensure that they are suitable for the job. Verify that the procedures have been approved and meet any jurisdictional requirements.
3. Verify that the current revision of each procedure is available at each welding station.
4. Witness any required performance tests prior to the start of fabrication such as bend tests, hardness tests, fissuring tests, etc. Many companies require the preparation of test welds by every welder who is to work on the project. These welds may be subject to bend tests, tensile tests, microstructural examination, etc., to establish the quality.

   In addition, corrosion testing of alloy coupons to meet ASTM A 262[2] or some other recommended practices may be a requirement and is usually performed prior to the start of fabrication as well as at various times throughout fabrication.
5. Confirm that the proper welding consumables are available and that they are stored correctly before fabrication begins. Welding rods are stored in sealed containers, while wire spools may be stored in boxes, cabinets, or on racks. Spools stored on racks require cleaning to remove dust, dirt, and other debris before they are placed in the welding machine. Low-hydrogen rods are stored in ovens at elevated temperatures to prevent moisture contamination. Electrodes with damaged, wet, or oily coatings are unacceptable.
6. Verify that rolls to be used during fabrication are clean and will operate satisfactorily.
7. For alloy welding, verify that the fabrication area is clean, properly isolated from contamination from other areas, and of adequate size for the fabrication.
8. If fabrication is to take place outdoors, ensure that the work will be properly sheltered from rain, wind, dirt, and other forms of contamination. Fabrication within 30 ft (10 m) of a roadway is not recommended, particularly in northern climates where deicing salts may be used in the winter. Check the temperature because fabrication at low temperatures requires preheating.
9. Examine rolls, shears, etc., to ensure that they are clean and will not create defects such as rolled-in scale, gouges, etc. For alloy fabrication, some manufacturers may use special rolls or they may place paper on the rolls to prevent iron contamination. A nitric acid test is often used after rolling to ensure freedom from iron contamination for high alloy material.
10. Weld preparations on plates are generally performed prior to rolling, although for thin material that is simply butted together, no preparation may be necessary. Testing (liquid penetrant, magnetic-particle, or ultrasonic techniques, as applicable) is used to reveal laminations, laps, or other defects on the weld preparations. Normally performed prior to plate rolling, the test may be carried out after rolling to reveal any damages that might have occurred from the rolling.

11. After rolling, the shell sections should be examined for mechanical damage (cracks, tears, gouging, etc.).

12. Prior to welding, the shell section diameters need to be measured to ensure that they conform to the specification.

13. Measure the joint fit-up and items such as the root gap, land area, and included angle to make certain that they comply with the approved welding procedures.

NOTE: See Appendix 5 for typical sketches.

14. Observe fit-up techniques such as tack welding or dogging. If excessive hammering is used to effect a fit-up, damage to the joint edges may result. Denting, flat spots, or excessive waviness are signs of damage that may result in cracking. Liquid penetrant or magnetic-particle examination may be required to check for cracking. Stress relieving of the damaged shell may be necessary. Heating with a torch and sledging edges into place is unacceptable.

15. Examine run-in and run-out tabs, if required. Determine that they are of the correct material and in the proper location. Tabs are often mounted too high or too low, creating a step at the weld end. This frequently results in a short end crack on the weld, usually in the center of the bead.

Inspection During Fabrication Welding. During fabrication welding, check for the following:

1. Observe the deposition of some of the root pass. Confirm that current and voltage settings are as outlined in the procedure.

2. Verify that the consumables used are correct. Low-hydrogen electrodes should only be removed from their heated container as required. Rods lying on equipment or on the floor are not acceptable.

3. Examine wire feed rates, shielding, and backing gas flow rates, as applicable. For submerged arc, ensure that the flux is flowing properly.

4. Observe rotation speed of the work or the travel speed of the welder to determine if the welding speed is as outlined in the procedure. A stop watch may be used to time a welding interval and then the inches (centimeters) travelled per minute can be determined.

5. If the underside of the root is back-gouged and weld metal-deposited, observe some of the grinding and the welding as well. Excessive grinding could induce cracking or result in undercut along the sides of the weld.

6. Observe the slag removal or other cleaning of weld deposits, as applicable. Grinding is not acceptable as a cleaning operation, as it may cause cracking and produce an undercut. Grinding may be required between passes with some alloys, however it must be closely controlled.

7. Following completion of the longitudinal welds, check the shell sections for out-of-roundness using a template, spring bars, or other suitable means. Short shell sections may be examined by placing them on a flat plate and using a beam compass, provided an accurate center point can be established. Conical sections are checked at both open ends and for flattening along the welds.

8. Examine both the inside and the outside of the finished welds. Ensure that all grease, dirt, scale, smut, etc., has been removed before testing. This is particularly important for liquid penetrant examination.

9. Examine both the inside and the outside of the welds for excessive reinforcement, weld edge extension, undercut, and surface defects such as porosity, cracks, center line shrinkage, weld spatter, included wire, icicles, burn through, and incomplete penetration.

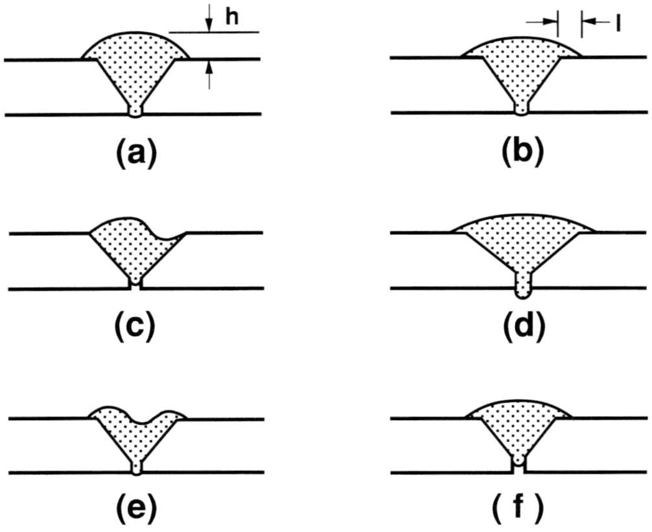

(a) Weld reinforcement (d) Excessive penetration
(b) Weld edge extension (e) Center-line shrinkage
(c) Undercut (f) Incomplete penetration

10. Ensure that the welder has applied his identification to each weld, either by stamping the component or attaching a metal tag containing the information.

11. During welding, ensure that other processes and work being carried out in the shop cannot contaminate the welds. Ensure that welding is properly shielded from drafts and rain water (from the outdoors, a leaking roof, open window, etc.).

12. Ensure that testing is not being performed on sections while welding is still underway on those parts, particularly if magnetic-particle examination is involved.

13. Ensure that longitudinal welds do not line up on successive shell sections. "Match-marking" shell sections may be required to ensure staggering of the welds (a minimum of 30° is general practice).

14. Ensure that welds are not placed on top of one another and that nozzles, manways, reinforcing pads, etc., do not interfere with major longitudinal or circumferential welds. If reinforcing pads are to cover some of the welds, make certain that the covered welds have been tested and accepted before covering.

15. Verify that a taper (3-to-1 minimum, but 4-to-1 preferred) has been used in joints between metals of differing thickness.

16. Ensure that nozzle cutouts are the right size and that weld preparations are tested, as required. Obtain samples of the cutouts for future reference and/or testing.

17. Ensure that corners on reinforcing pads, lugs, saddles, etc., are rounded to reduce stresses.

18. Check that nozzles are installed in the correct location with reference to an established centerline or reference line. Angular displacement is usually checked with a protractor or compass. A convenient reference line often used is one through the center of the manhole or the nameplate.

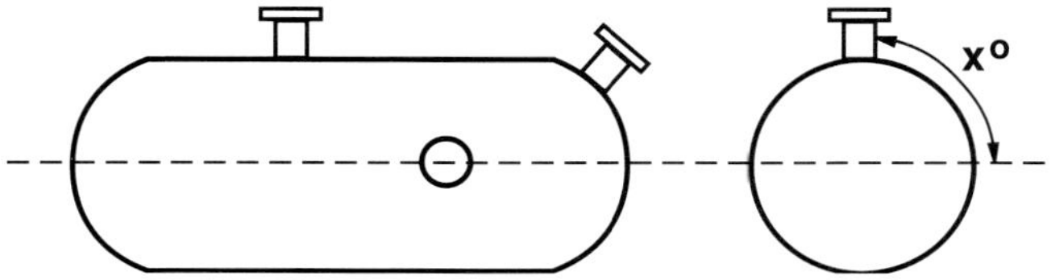

19. With a square, measure the length of the nozzles above the shell. Ensure that they are either at right angles to the shell or at other specified angles.

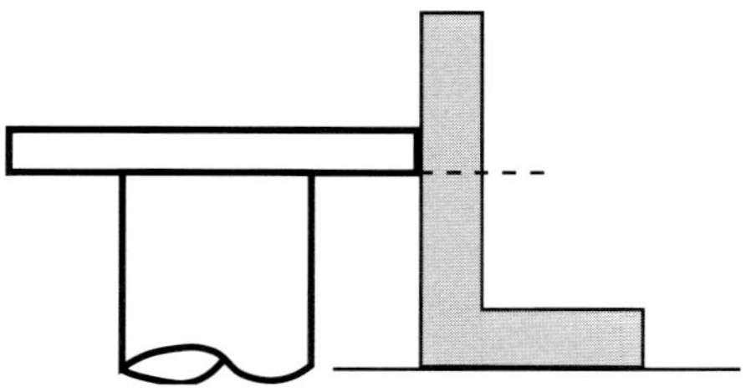

20. Using a square or level, make certain that the flanges are square; not tilted.

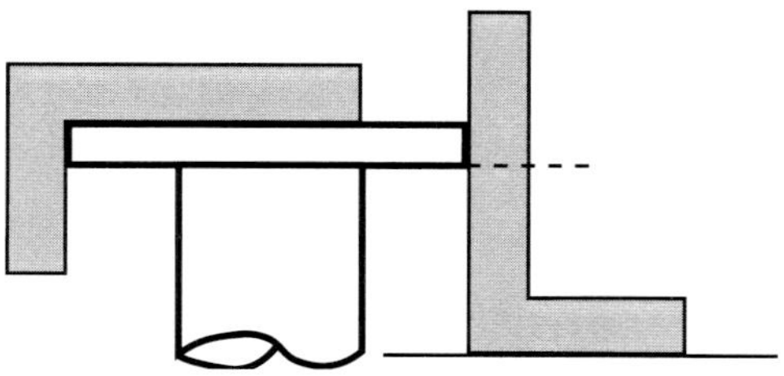

21. Witness stress relief, as applicable. Ensure that thermocouples are in good condition and are reading the temperature of the component. Review a copy of the heat treat chart to verify compliance with the specification. Sign and date the chart.

22. Check to be sure that all reinforcing pads contain threaded weep holes and ensure that they are not leaking during hydrostatic tests.

Testing During Fabrication. The following are items one should be aware of in regards to testing.

1. Witness some NDT, as applicable. Review radiographs, as applicable.
2. For spot radiography, select spots to:
   a) test each longitudinal and circumferential seam;
   b) check the welds of every welder who welded on the vessel;
   c) check welds for all techniques used on the job; and
   d) check areas of high stress, such as at T-junctions.

NOTE: It is also advisable to check spots on either side of T-junctions, as welders may be more careful at the T's. Variations of spot selection helps to ensure overall quality. If additional spot tests are required because of rejection, ensure that a spot is selected on either side of the previous failed one. At T-junctions, this normally involves at least one spot on the vertical leg and one on each arm of the T.

3. Mark all spots requiring repairs or verify their marking.
4. Ensure that defects requiring repairs be located on a defect map for future reference.
5. Witness some defect removal along with the testing that confirms its removal, i.e., liquid penetrant, magnetic particle, ultrasonics, etc.
6. Witness some of the repairs and subsequent re-testing.
7. Examine the inside of the vessel to ensure that nozzles are welded on the inside and that brackets, tray bars, and/or other internals are in place at the correct locations.
8. Verify that tray supports and/or other internal supporting devices are installed so that the trays, screens, packing, etc., will lie at the appropriate angle. This is often accomplished by using a steel straightedge and a level across the support at 0°, 45°, 90°, and 135° to confirm flatness.

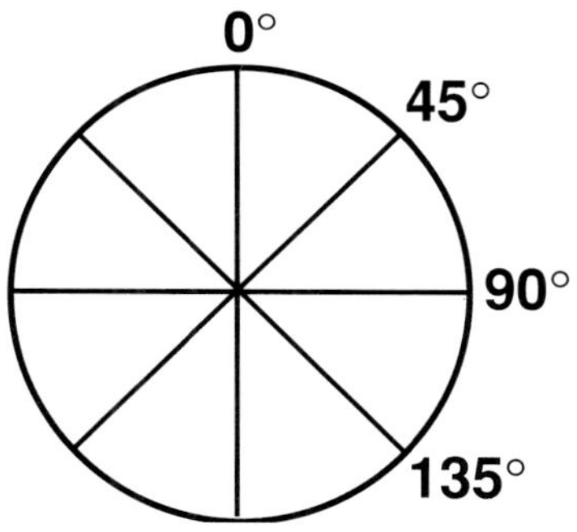

If the vessel is fabricated on its side, place the straightedge across it in the vertical position. Rotate 45°, repeating the prodecure until all four angular displacements have been verified. For internals that must display a slope, a straightedge is again employed, but with an angular level. This level is adjusted to yield the specified slope and then placed on the straightedge to determine whether or not the slope in the vessel conforms to specification.

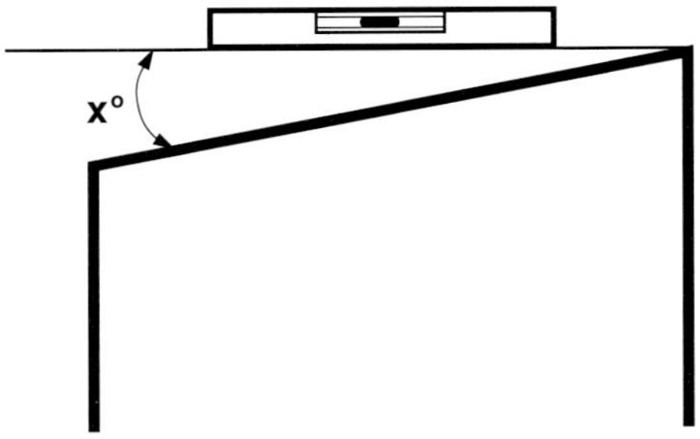

9. After all welding (including repairs) has been completed, examine the shell for any signs of distortion such as bulges or dents, flat spots, peaks in the welds, waviness, gouges, and tears. Pay particular attention to thick/thin areas, angular changes, skirts, legs or saddle areas, nozzle locations, and head to shell welds. Also examine carefully areas where support bracing may have been welded in during fabrication and the locations for internal baffles, supports, and other structures.

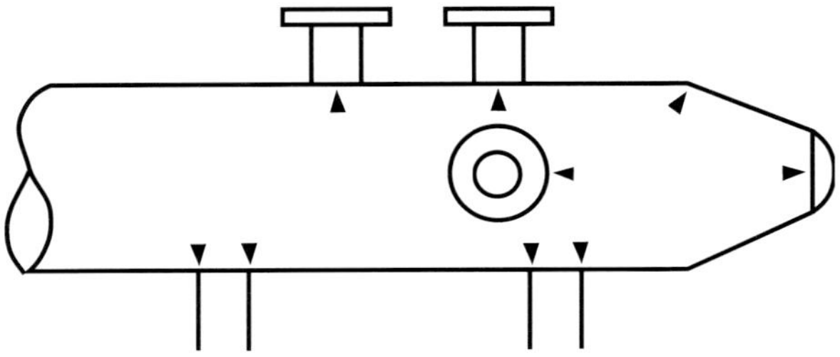

10. Perform a complete dimensional check on the finished vessel. This check is performed after all postweld heat treatment has been completed. It may be performed before or after the hydrostatic test, as desired.

11. Witness the hydrostatic test ensuring that the pressure reached and held is that required by the applicable code; usually one and one half times the design. The pressure is held for a minimum of one-half hour. Check the pump and pressure gauge to ensure that they are indeed fitted to the vessel being tested. For example, request that the vent be opened, then watch to see if the pressure drops. The vent may then be closed and the test continued.

    CAUTION: Do not approach the vessel until the test pressure has been attained and is holding. Also, to avoid the possibility of brittle fracture in carbon and low alloy steel components, the testing water is maintained at a minimum temperature of 60°F (16°C).

12. During the hydrostatic test, examine all flanges, couplings, weld seams, and reinforcing pad weep holes for signs of leakage. After this examination, check to be sure that the gauge pressure has not "dropped-off" and that the pump has not been "run" so as to maintain the pressure.

13. Following the hydrostatic test, ensure that the vessel is thoroughly drained and dried to prevent corrosion (or even freezing damage) during shipment and storage. Austenitic stainless steels may be particularly susceptible to microbiologically influenced corrosion if the equipment is not properly drained and dried.

14. After the hydrostatic test has been completed, examine the vessel for such signs of distortion as bulging, swelling, sagging, and tears in support welds.

Inspection of Stress Relieving

For vessels requiring stress relief, it is essential to conduct a complete visual and dimensional inspection prior to the stress relief. Vessel internals such as trays, screens, etc., are not installed prior to the stress relief.

Stress relieving is carried out following the completion of all welding on the vessel and before final acceptance testing. There is some difference of opinion as to whether stress relieving should be performed before or after NDT, e.g., radiography. Some companies argue that they do not want to incur the cost of a stress relief only to find that they must repeat it if post stress-relief radiography indicates repairs are necessary. Purchasers claim, quite legitimately, that stress relieving may induce defects. Therefore, they require that final NDT be performed only after the stress relief is complete. The hydrostatic test is always performed after stress relieving. A preliminary leak test prior to stress relief might be advisable.

Stress relieving inspection generally consists of the following steps:

1. Verification that the correct stress-relieving procedure is available to the furnace operator and that the temperatures specified are correct for the vessel's materials. Heating temperatures for carbon steels are frequently limited to a maximum of 200°F (93°C) per hour (ASME B & PV Code is 400°F/h/in. [200°C/h/2.5 cm]). See "Pressure Vessels: The ASME Code Simplified" in the Bibliography (Appendix 8). Lower or even higher rates may be specified for other alloys or for vessels with complicated shapes. Heating and cooling requirements are outlined in the ASME Code, paragraphs UCS-56 and UHA-32.

2. Verification of the soak time, usually a minimum of fifteen minutes for very thin material; one hour per inch (2.5 cm) of thickness for carbon and low alloy steels—less than five inches (13 cm) thick; or five hours plus fifteen minutes for each additional inch for material over five inches (13 cm).

3. Verification of the initial charge temperature. This depends on the alloy, but rarely exceeds 400° to 600°F (190° to 300°C), although the ASME Code permits 800°F (426°C).

4. Verification of the cooling method to be used and the rate of cooling. If quenching is involved, check that the medium is compatible with the alloy being heat-treated and that quenching can be accomplished without distortion.

5. Ensure that any dissimilar metals will not be affected adversely by the heat treatment.

6. Confirmation that both the fuel and the furnace environment are compatible with the materials, thus ensuring that they will not create high-temperature corrosion problems.

7. Verification that the recording chart is operational and that thermocouples are actually attached to the appropriate recording devices.

8. Verification that the thermocouples are functioning properly.

9. Check the thermocouple locations and make certain that they are securely attached to the vessel. Verify that the thermocouple locations will provide an adequate indication of the temperature in the entire vessel. Verify that there are no "dead spots" or baffling that will prevent maintaining an even temperature throughout the furnace.

10. Verify that the vessel is adequately supported or blocked to prevent sagging, buckling, or complete collapse.

11. Confirm that flange faces, threaded openings, holes, and other attachments are adequately protected.

12. Verify that there will be no flame impingement on the vessel.

13. Check the means of controlling the temperatures to ensure the appropriate heating, soaking, and cooling operations are being carried out.

14. Sign and date the heat-treat chart at the completion of the operation.

15. Following heat treating, examine the vessel, particularly in the welds, flanges, and support areas for any sign of distortion, cracking, tears, etc. Check the vessel for out-of-roundness, sagging or bent nozzles, cracks or tears at

supports, severe oxidation, and other defects. Verify that internal support structures have not shifted.

16. Ensure that the vessel is cleaned properly before the hydrostatic test.
17. Ensure that scale and dirt are not plugging weep holes in reinforcing pads.
18. Examine flange faces for signs of damage.

Sectional Stress Relieving. For vessels that are too long to fit in the furnace for a single stress relief operation, stress relieving may be carried out in two or more stages. In these particular cases, a portion of the vessel will project out into the shop and, therefore, remain cooler. There may be a point where undesirable metallurgical structures are generated (for instance, 885°F [437.5°C] embrittlement or carbide precipitation in susceptible alloys); therefore, it is essential to ensure that these areas are properly heat treated to prevent premature failure. This is usually accomplished by dividing the vessel into sections and then stress relieving the sections, ensuring an adequate overlap, usually a minimum of 5 ft (1.5 m). The portion of the vessel outside the furnace is insulated or shielded so that the temperature gradient does not damage the structure.

Vessel sections or welds may also be locally stress-relieved. This involves heating a band or area around the entire vessel. The width of the band will depend upon the method of application and whether or not nozzles are involved. Without nozzles, the minimum width of the band on each side of the greatest width of the weld is twice the shell thickness. If nozzles are involved, the minimum width of the band is equal to six times the plate thickness beyond the welding attaching the nozzles to the shell and extends outside any reinforcing pads.

Heating bands around the vessel may be accomplished using portable burners, electrical elements, or other controlled means. Insulation is usually placed on one side of the shell and heat on the other. If permanent insulation is used, care must be exercised to prevent damage and to ensure that the insulation is adequate for the operation.

Inspection of vessels heated in more than one operation requires the following additional steps:

- ensuring that there is adequate support for the portion of the vessel that is to be outside of the furnace;
- monitoring the temperature on the portion outside of the furnace to ensure that there is no harmful gradient;
- ensuring an adequate overlap; minimum 5 ft (1.5 m) for successive sections;
- ensuring that thermocouples are attached to each section of the vessel to be stress-relieved;
- examining and signing heat-treat charts for each section; and
- examining the vessel for distortion, particularly in the overlap regions.

Hardness tests may be advisable in carbon and low-alloy steels to ensure no degradation occurs in the regions of overlap. With austenitic and other high-alloy materials, coupons are frequently heat-treated along with the vessel in the overlap regions. These are subsequently examined microscopically for signs of structural damage (primarily carbide precipitation). They may also be subjected to intergranular corrosion tests.

NOTE: To determine an adequate overlap for a vessel, consider both the furnace and vessel length. The maximum length of a vessel that can be accommodated in a furnace that opens only at one end is twice the furnace length minus 5 ft (1.5 m). For example, if the furnace length is 30 ft (9.1 m), then a 55 ft (16.8 m) long vessel could be stress-relieved in two heats providing a 5 ft (1.5 m) overlap.

Longer vessels are often stress-relieved in furnaces that open at both ends. The vessel is passed through the furnace in stages, with both temperature gradient and shielding being monitored at both ends.

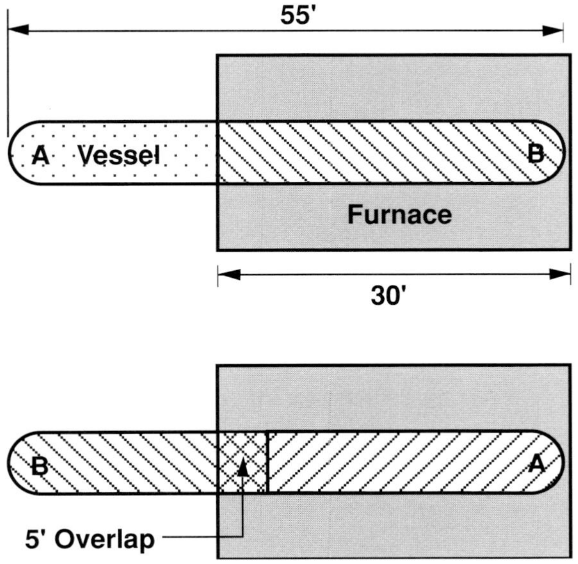

Still other vessels may be furnace stress-relieved in sections as they are built. Then the circumferential joints are locally stress-relieved, provided the longitudinal welds in the central sections will not be seriously affected by such treatment.

Band Stress Relieving. When stress relieving using bands around an entire vessel, great care must be exercised to prevent distortion and harmful temperature gradients. This is accomplished most easily by the following:

1. Ensuring that the band is of adequate width to produce a proper stress relief;
2. Monitoring the temperature on thermocouples placed around the shell (minimum top, bottom, and half-way in between).

NOTE: On sections with nozzles, at least one thermocouple is placed in the nozzle region.

3. Monitoring the temperature outside the heated band.
4. Ensuring that proper heating and cooling sequences have been followed.
5. After stress relieving, the vessel is inspected for such distortion as sagging, buckling, and flat or high spots in the vicinity of the stress relief (in an area at least the width of the stress relief band on each side of that band).
6. Requiring signed and dated recording charts. Producing a chart for each band. Some compa-

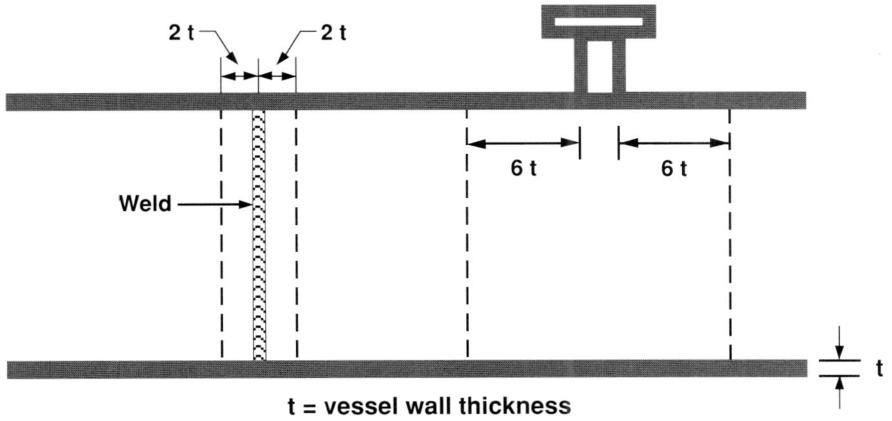

BAND STRESS RELIEF

nies may attempt to use one chart for several bands that are heat-treated at the same time; this is unacceptable.

Stress Relieving of Multiple Components. Stress relieving of a number of components may be performed at one time. Components are usually stacked or piled in the furnace and the whole charge is heated. Thermocouples are not placed on each component but are placed to provide temperature readings at the bottom, top, and in the center of the stack or pile. Additional thermocouples may be used in other areas or zones where temperature variations are anticipated.

Stress Relieving of Different Thicknesses. Vessels or components with differing thicknesses may be stress-relieved in the same batch. In this case, the heating, soaking, and cooling sequence used is that required for the thickest component. Consequently, it is necessary to ensure that such a heat treatment will not damage the thinner materials.

Stress Relieving at Lower Than Specified Temperatures. Stress relieving of components, particularly of low carbon steels, may be accomplished by heating the vessels at lower than specified temperatures for an extended period. The current ASME Code, Section VIII, Division I, lists the following guidelines for carbon and low alloy steels:

Decrease in temp below normal holding temp.	Minimum holding time at decreased temp. in hours/inch of thickness
50°F (10°C)	2 h
100°F (38°C)	3 h
150°F (66°C)	5 h
200°F (93°C)	10 h

However, as these recommendations may change, it is wise to check the latest edition of the Code, especially if lower-than-normal temperatures are to be used. In addition, the lower temperature/longer time stress relief may not be as effective as a full-temperature treatment.

Documentation

Required documentation includes the following:

1. Ensure that the affidavit of manufacture has been completed by authorized inspectors.
2. Check the vessel nameplate to verify that the correct data has been included. A rubbing or photograph of the nameplate may be obtained for inclusion in a history docket.
3. Examine the documentation assembled as a history docket for the vessel. The following information is generally included as a minimum.

    a) An as-built drawing showing the vessel design; stamped, signed, certified, or approved by a professional engineer as per jurisdictional requirements.

    b) Mill test certificates, certificates of compliance, or other suitable documents outlining material chemistry and properties.

    c) All test reports such as radiographic, ultrasonic, magnetic particle, hardness, and the qualifications of testing personnel.

    d) The welding procedures used in the fabrication.

    e) Welder certification for each welder used on the vessel, along with welder identification numbers, weld maps, or other methods of identification, as applicable.

    f) Heat-treat charts, if applicable.

    g) Hydrostatic test reports.

    h) Affidavit of manufacture.

    i) A shipping release signed by the purchaser's inspector outlining the condition of the vessel prior to shipment. This release, if required, is issued after the hydrostatic test if the vessel is a noncritical piece of equipment.

4. For critical vessels, witness the loading of the vessel and inspect all nozzles, flanges, etc., for signs of damage before issuing a shipping release.
5. Witness the off-loading of a critical vessel at the plant site and examine the vessel carefully for shipping damage.

Check Lists

The check lists that follow may aid the inspector. Since the lists are quite detailed, all items will not be applicable in every case. The inspector can select those details required for the examination of specific equipment and delete those that are not applicable.

NOTE: See Appendix 3 for heat exchanger check sheets used by a major chemical company.

PRESSURE VESSEL CHECK LIST

VESSEL NO.: UNIT NO.: DATE:
FABRICATOR: LOCATION: INSPECTOR:

RAW MATERIAL					DOCUMENTATION			
ITEM	SPEC.	I.D.	THICK.	COMMENTS	ITEM	ACCEPT	REJECT	COMMENTS
HEADS					DRAWING			
NOZZLES					MILL CERTIFICATIONS			
FLANGES					NDE TEST REPORTS			
COUPLINGS					HYDROSTATIC TEST			
SHELL PLATE					PURCHASE ORDER			
BOLTS					SPECIFICATIONS			
NUTS					FABRICATION PROCEDURES			
REINFORCING					QA MANUAL			
VALVES					AFFIDAVIT OF MANUFACTURE			
NIPPLES					NON-CONFORMANCE RESOLUTION			
TUBES								
PIPING								

INSPECTION CHECK LIST

RAW MATERIAL				NDE AND OTHER TESTS			
ITEM	ACCEPT	REJECT	COMMENTS	METHOD	ACCEPT	REJECT	COMMENTS
WELDING PROCED.				VISUAL			
WELDER QUALIF.				RADIOGRAPHY			
JOINT FIT-UP				ULTRASONICS			
TACK WELDS				EDDY CURRENT			
RUN-IN / RUN-OUT				MAG. PART.			
DISTORTION CONT.				LIQUID PENETRATION			
CONSUMABLES				HYDROSTATIC			
WELD TECHNIQUE				TEST / PRESSURE			
SLAG REMOVAL				LEAK TEST			
WELD REINFORCE				PNEUMATIC TEST			
WELD QUALITY				IMPACT			
WELD SHIELDING				HARDNESS			
				CORROSION			
				TENSILE			

DIMENSIONS					
ITEM	**SPEC.**	**ACTUAL**	**ACCEPT**	**REJECT**	**COMMENTS**
EXCHANGER TUBING • DIAMETER • THICKNESS • LENGTH • DEPTH OF ROLL • SEALWELD **TUBESHEETS** • THICKNESS • PITCH OF HOLES • CLADDING • HOLE DIAMETER • FLATNESS					

MISCELLANEOUS					
ITEM	**SPEC.**	**ACTUAL**	**ACCEPT**	**REJECT**	**COMMENTS**
INTERNALS • TRAYS • BAFFLES • AGITATORS • MIXERS • PACKING • CATALYST • LININGS **HEAT TREAT.** • STRESS RELIEF • ANNEAL • NORMALIZE **SURFACE PREP.** • GRINDING • BLAST CLEANING • PICKLING • PHOSPHATE					

DIMENSIONS					
ITEM	SPEC.	ACTUAL	ACCEPT.	REJECT	COMMENTS
LENGTH					
DIAMETER					
MINIMUM THICKNESS					
• SHELL					
• HEADS					
• NOZZLES					
OVALITY					
NOZZLES					
• DIAMETER					
• NUMBER					
• LOCATION					
• ORIENTATION					
• ALIGNMENT					
• REINFORCEMENT					
FLANGES					
• SIZES					
• TYPE					
• BOLT HOLES					
SKIRT					
• DIAMETER					
• THICKNESS					
• MANWAY					
• BOLTING					
BOLTS					
• TYPES					
• SIZE					
• THREADED INTO NUTS					

MISCELLANEOUS					
ITEM	SPEC.	ACTUAL	ACCEPT	REJECT	COMMENTS
PAINTING					
COATING					
LINING					
INSULATION					
OIL					
GREASE					
PACKAGING					
LOADING					
SHIPPING DAMAGE					
NAMEPLATE DATA:					
SPECIAL INSTRUCTIONS:					

CHAPTER 7

Piping

General Overview

Pipe can be made from any ferrous or nonferrous material or alloy that can be drawn through dies, rolled and welded, or cast to form tubular sections. Therefore, pipe may be seamless such as that produced by drawing, extrusion, or casting, or it may be welded. Welded pipe may be produced with a single longitudinal seam or it may have a spiral seam (a weld that spirals around the pipe as it proceeds along the length).

All pipe, regardless of the method of manufacture, can be produced to a variety of diameters and thicknesses.

Following welding, pipe may be drawn to improve its properties or it may be drawn and heat-treated. Rarely is welded pipe used in the as-welded condition.

The manufacture of pipe in nominal sizes, from $^1/_8$ in. (3 mm) up to 12 in. (30 cm), inclusive, is based on a standard OD. Originally, this OD was selected so that pipe having the standard OD and a typical thickness would have an ID approximately equal to the nominal size, i.e., 8-in. pipe would have an 8-in. (20-cm) ID. For piping over 12-in. (30-cm) nominal size, the size denotes the actual OD, i.e., 14-in. pipe has a 14-in. (35-cm) OD.

ASTM specifications use NPS (Nominal Pipe Size) which is a *dimensionless* designation. These designators relate to nominal pipe sizes in inches. For example, NPS 8 is nominal 8-inch pipe, NPS 12 is nominal 12-inch pipe; etc.

Wall thickness is indicated in one of two ways. The system employing the designations "Standard," "Extra Strong" (XS), and "Double Extra Strong" (XXS), has been used for many years. When schedule numbers were developed, there was resistance to replacing terminology that had been in general use for many years. As a result, piping may be ordered by schedule number or by the Standard, XS, and XXS designations.

Standard and Schedule 40 are identical for nominal pipe sizes up to and including 10 in. (25 cm). All larger sizes of Standard wall have a $^3/_8$-in. (9.5-mm) wall thickness. Extra Strong and Schedule 80 are identical for nominal pipe sizes up to and including 8 in. (20 cm). All larger sizes of XS Wall have a $^1/_2$ -in. (12.7-mm) wall thickness. There is no correlation between XXS and Schedule numbers except for 10-in. (25-cm) and 12-in. (30-cm) sizes. Double Extra Strong is equivalent to Schedule 140 and Schedule 120, respectively. Double Extra Strong is not used above the 12-in. (30-cm) size. See Appendix 4 for tables of thickness.

The thickness of drawn or cast pipe may vary from the nominal by a manufacturing tolerance which may be as much as $\pm$ 12.5%. For critical or hazardous service, many companies specify a minimum wall thickness so that the tolerance is $\pm$ 12.5% to ensure sufficient thickness for the process. This is particularly desirable on fittings that may be thinned during forming operations.

The chemical, petrochemical, and refining industries generally use seamless pipe for *severe* service, although carefully inspected welded pipe is used successfully. In carbon steel and low-alloy material, seamless pipe is produced in sizes up to

26-in. (66-cm) OD, but common practice seems to limit seamless material to 12-in. (30-cm) OD. With austenitic stainless steels and other high alloys, seamless pipe appears to be limited to a maximum of 12-in. (30-cm) OD, although some manufacturers do not produce pipe with more than an 8-in. (20-cm) OD.

Joining Methods

Piping may be joined in several ways, but the most common methods are bolting, threading, and welding. Flanged or threaded connections are used to join pipe to equipment, valves, etc., if the connections must be opened or broken for maintenance, draining, or other operations.

Bolted (Flanged) Joints. Bolted pipe joints generally indicate the fastening together of two flanges using bolts, studs, or some other approved threaded fasteners with some form of gasket between the flange seating surfaces. The gasket may be flat with a texture ranging from serrated to smooth, depending on the type of gasket, the gasket material, and the service conditions; or it may be in the shape of a ring to fit into machined grooves in the flange face. Common flange facings are raised face, ring-type, and flat face.

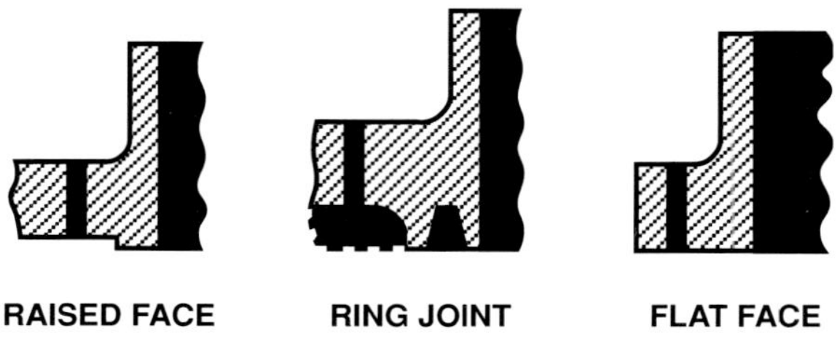

Flanges are commonly weld neck, slip-on, threaded, blind, lap joints, or socket welding types.

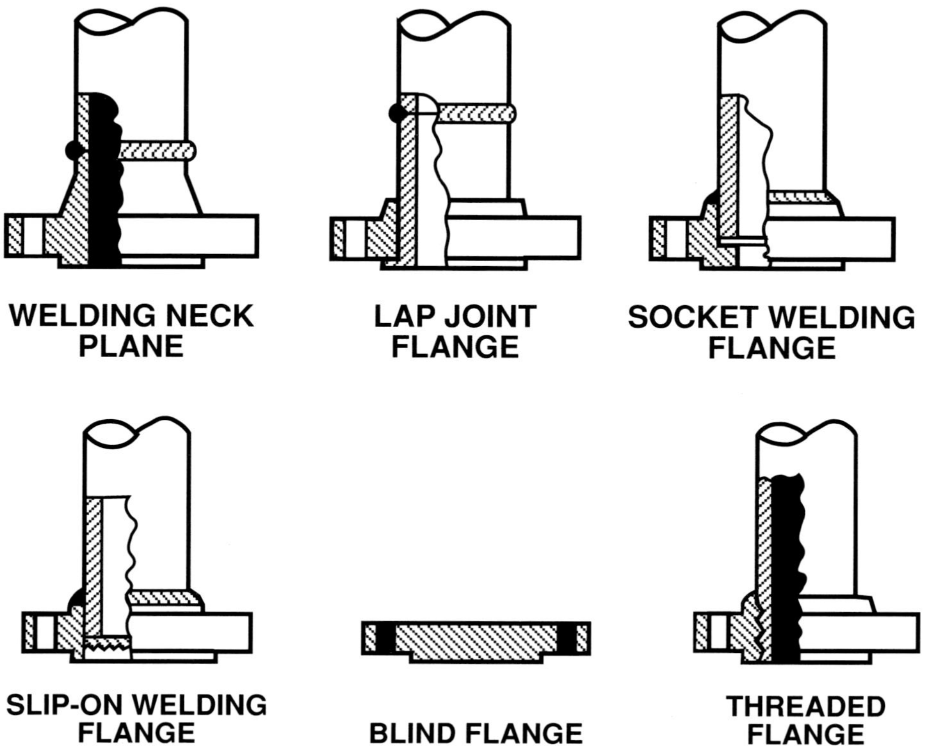

Threaded Connections. Threaded connections on pipe 24 in. (60 cm) and smaller are outlined in the ANSI[4] /ASME Standard (B1.20.1).[3] However, for pressure piping, threaded connections are generally limited in use to pipe 2 in. (50 mm) and smaller. For service that may involve hydrogen or may generate hydrogen because of corrosion, threaded connections may be prohibited. For other critical service, threaded joints may be seal-welded ensuring that all threads are covered.

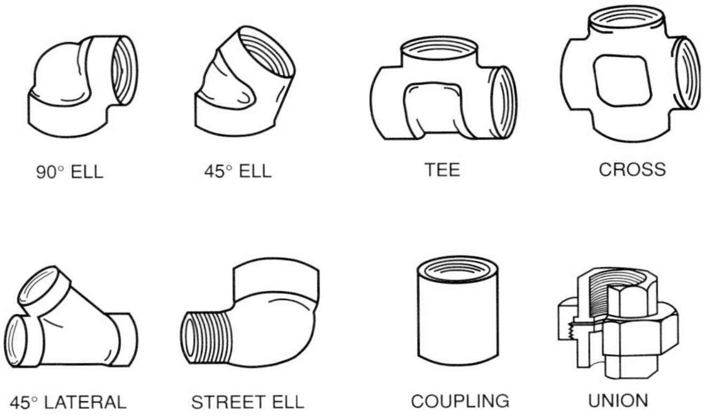

Welded Connections. Welded joints have largely replaced threaded and flanged connections as a method of permanently joining sections of pipe together. By eliminating threaded and flanged joints, potential leak areas may be reduced.

Joints can be butt-welded for pipe of any size, but socket welding is often employed on piping 2 in. (50 mm) and smaller. Butt welds are produced by shaping the ends of the pipe, fitting, valve, etc., by first grinding or machining, then butting the adjoining ends, and finally welding the joint with Shielded Metal Arc Welding (SMAW), Gas Metal Arc Welding (GMAW), Gas Tungsten Arc Welding (GTAW), or some other acceptable technique. Welding is performed to an approved welding procedure in a manner similar to that for pressure vessels (Chapter 8).

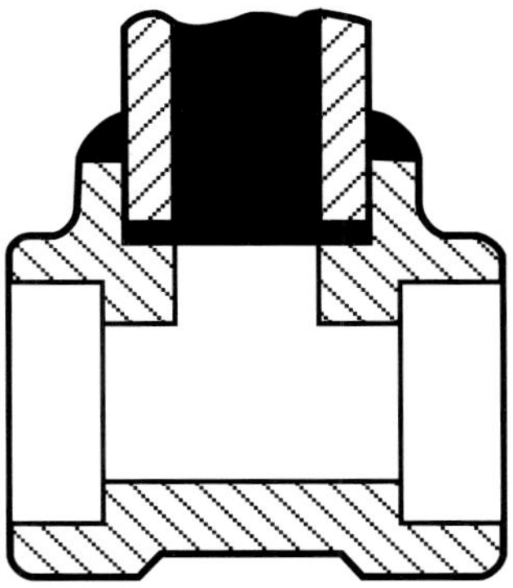

SOCKET-WELDED TEE CONNECTION (Cross Section)

Socket-welded connections are made by inserting the end of the pipe into the fitting or valve and then fillet welding the joint. A space must be provided between the pipe end and the bottom of the socket to allow for expansion. Specifications for socket welding generally recommend first inserting the pipe end fully into the fitting or valve, backing it out $1/8$ to $1/4$ in. (3 to 6 mm), and then welding. This method ensures a gap for expansion.

Other Types of Joints. There are a number of proprietary joints available that incorporate unique designs of gaskets, clamps, and bolting arrangements. Proprietary joints can offer certain advantages over conventional joints in specific services. Some permit "quick" pipe disconnection, which requires frequent change such as loading lines.

In addition, conventional joints are generally not used with cast iron pipe, particularly if threading or welding is involved.

Cast iron pipe joints are generally flanged, bell-and-spigoted, or packed-and-sleeved. Push-on joints employing rubber or synthetic ring gaskets are also available.

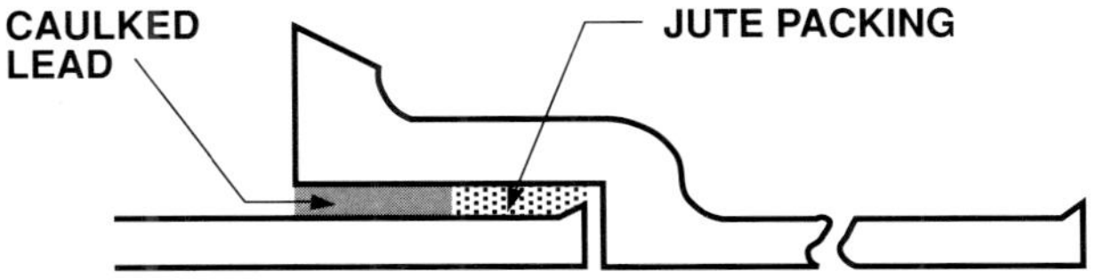

BELL-AND-SPIGOT JOINT (Cross Section)

---

[4] American National Standards Institute, New York, NY.

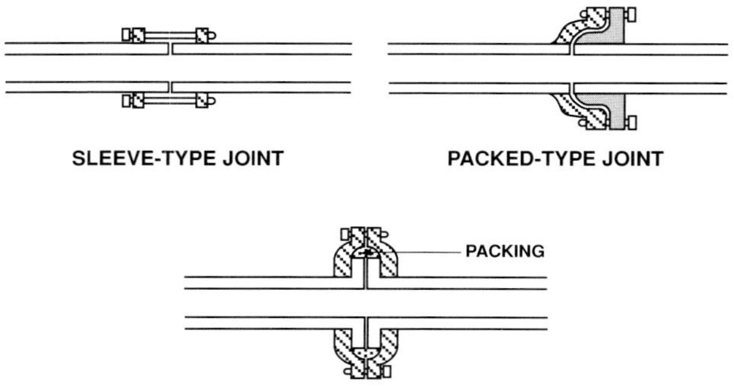

SLEEVE- AND PACKED-TYPE JOINTS
(Cross Sections)

Inspection of New Shop Fabricated Pressure Piping

Piping to be used in the chemical, petrochemical, refining, and similar industries is generally inspected at the fabricator's mill, in the supplier's storage yards, or in the shops where spool pieces are to be fabricated. Most companies order pressure piping to the requirements of ANSI/ASME B31.1[4] and B31.3,[5] etc., as applicable and/or to ANSI/ASME B31.2,[6] API 5CT,[(5), 7] 5L,[8] etc., as applicable. Consequently, it is important for the inspector to confirm that the piping offered for inspection has been manufactured to the required specifications.

Pipe is usually classified as seamless (S); electric resistance-welded (E) or [ERW]; fusion-welded (F) or [FW]; and butt-welded (B) or [BW]. The letters in parentheses are API designations, while those in brackets are used by some overseas mills or requested by some purchasers to permit more positive identification of components. Each class of piping may contain several types of pipe made by different processes, i.e., seamless may be made by extrusion, piercing, drilling, punching, forging, and casting.

Although there are many different types of pipe made by various processes, the inspection requirements are common to all.

General Inspection. The following requirements pertain to all types of pipe.

1. Verify that the piping has been produced to the purchase specifications and is marked accordingly. The pipe may be marked by dye stamping, paint, stenciling, or rolling with:

   a) the name and decal of the manufacturer;

   b) the kind of pipe, i.e., seamless, electric resistance-welded, etc.;

   c) a thickness designation, generally the schedule number, but Standard, XS, or XXS may also be used; and

   d) the specification number and grade, i.e., SA 106 Gr. B, API 5L, etc.

Additional requirements, depending on specifications, may include:

- the length of the pipe;
- the hydrostatic test pressure, as applicable;
- if the pipe is not hydrostatically tested it is marked (NH); and
- pipe that has been weld-repaired may be marked (WR).

Marking generally begins about 12 in. (30 cm) from the end of each length of pipe. For small diameter pipe, usually less than 1 to 1½ in. (2.5 to 3.8 cm), or for short lengths less than 3 ft (0.91 m) in length, the required data may be outlined on a tag securely fastened to the bundle or box in which the pipe is shipped.

2. Examine mill test reports to ensure that the chemical and physical properties are acceptable for the material specified. If inspection is carried out at the mill, verification of ladle chemical analysis may be performed. In addition, tensile tests, flattening tests, impact tests, etc., may be witnessed.

---

(5) API 5CT (US Customary Units) and API 5CTM (Metric Units) have replaced API 5A, "Specification for Casing, Tubing, and Drill Pipe."

3. Periodically examine pressure gauges to ensure that they are in good condition and have been calibrated within the past six months. Gauges that have not been calibrated may be reading incorrectly.

4. Verify the results of any on-line NDT such as electromagnetic, ultrasonic, and radiography.

5. Examine the pipe for visible surface defects. This is usually done by examining lengths selected at random on large orders or by examining the entire lot for small, critical lots. Do not attempt to examine pipe in bundles or in storage racks. Request that the manufacturer spread out the materials to be examined on a table, concrete pad, or some other suitable surface.

6. Examine the pipes surface finish ensuring that it is free of cracks, dents, gouges, laps, laminations, etc., on both the inside and outside surfaces. Rotate the pipe to examine all surfaces. Pay particular attention to the pipe ends as defects may be observed on such machined surfaces.

   Check the inside of the pipe by placing a light at one end and sighting down the pipe from the other end.

7. Require that heavy scale, dirt, rust, grease, oil, etc., be removed to permit examination of the underlying surface.

8. Surface defects may be removed by grinding as long as the depth does not encroach on the minimum specified thickness. If the grinding penetrates the minimum thickness, repair welding, if permitted by the specification, may be required. Defects not requiring repair are generally "feathered" to present a smooth contour.

9. All repair welds are inspected to the same requirements as the original pipe.

10. Observe any NDT not performed on-line such as ultrasonics, radiography, magnetic particle, and/or liquid penetrant. Pipe is usually divided into sections for testing. Pipe 18 in. (46 cm) and larger is divided into quarters, pipe 8 to 18 in. (20 to 46 cm) is divided into thirds, and for smaller piping only half of the pipe circumference is examined at a time.

11. Check the thickness of the pipe. This is done by measuring at each end with calipers, micrometers, or some other acceptable instrument. The pipe is measured at a minimum of three locations, spaced equally around the circumference. On pipe over 18 in. (46 cm), four locations around the circumference are selected.

    The wall thickness may vary depending on the specification, but in general it should meet the following requirements:

    - a thickness medium of 12.5% under the nominal wall for seamless and welded pipe with no filler;
    - for forged and bored pipe, the wall thickness minimum is the specified thickness (it may vary over the specified by a maximum of 1/8 in. [3.2 mm]);
    - for cast pipe, the specified thickness, but may vary over the specified by a maximum of 1/16 in. (1.6 mm);
    - on welded pipe with filler added, the thickness tolerance, although decided by individual specifications, is usually ± 10 to 15% (some companies prefer to specify a minimum thickness, producing a tolerance of +10 to 15%-0%).

12. Check the length of the pipe using a steel rule. Pipe may be supplied in definite cut lengths or in random lengths. For definite cut lengths, no pipe is to be supplied under the length specified but may be supplied over the length as follows:

    a) for seamless and welded pipe with no filler, the length may exceed that specified by not more than $^1/_4$ in. (6.4 mm); and

b) for all others, the length may exceed that specified by not more than $^1/_8$ in. (3.2 mm).

13. Check the ID using calipers or suitable gauges. The ID of most pipe may vary under by a maximum of $^1/_{16}$ in. (1.6 mm), but not over the specified requirement.

14. Check the OD. Outside diameter variations are usually based on pipe dimensions with the permissible over specified values increasing with pipe size as follows:

Pipe Size		Permissible variation over specified	
$1^1/_5$ in. or less	(3.0 cm or less)	$^1/_{64}$ in.	(0.4 mm)
$1^1/_2$ to 4 in.	(3.8 to 10 cm)	$^1/_{32}$ in.	(0.8 mm)
4 to 8 in.	(10 to 20 cm)	$^1/_{16}$ in.	(1.6 mm)
8 to 18 in.	(20 to 46 cm)	$^3/_{32}$ in.	(2.4 mm)
18 to 26 in.	(46 to 66 cm)	$^1/_8$ in.	(3.2 mm)

NOTE: Significant ovality may develop during annealing, straightening, etc., on thin wall pipe, i.e., pipe with a wall thickness of 3% or less of the OD. For pipe NPS 6 (6 $^5/_8$ in. [17 cm] and over), the ovality allowance is generally 1.5% of the specified OD.

15. Check the straightness by placing a 10 ft (3 m) straightedge along the pipe so that both ends are in contact with the pipe. The maximum permitted deflection is usually $^1/_8$ in. (3.2 mm).

16. Check the end finish. Check bevelled edges for the angle of bevel, the presence and width of the land, and also the squareness of the land. Check plain end pipe for the squareness of the end cut. Examine the pipe ends for gouges, nicks, visible laminations, etc. Witness any magnetic-particle or liquid penetrant tests performed on the pipe ends.

17. Verify that any heat treatment required has been performed and that charts are available showing the maximum temperature reached and the time at temperature.

18. If quenching is involved, verify that the quenching medium is correct for the alloy. Witness some of the quenching.

19. If required by the specification, verify the weight of the pipe. Standard pipe sizes are verified to ANSI/ASME B36.10[9] and B36.19.[10] For nonstandard sizes, the weight per foot may be calculated as:

$$W = 10.68\,(D-t)\,t$$

where: **W** = weight in lb/ft,
D = specified or calculated diameter,
t = specified wall thickness.

NOTE: Values quoted in ANSI/ASME and those determined by the above formula are for carbon steel pipe. Ferritic stainless steels may be about five percent less and austenitic alloys about two percent greater than these values.

Welded Pipe. In addition to the general requirements outlined above, welded pipe is inspected for the weld quality as follows.

1. Examine the weld procedure to ensure that it will produce welds meeting the specifications.

2. Examine skelp or plate edges for signs of laminations, laps, or offsets prior to welding. Ensure that any bevels are correct and that excessive tacking has not been used. Ensure that tack welds are free of cracks and other defects.

3. Examine the welds to detect partially welded or open seams. On large diameter pipe, travel down the inside of the pipe in order to examine the inside surface of the weld. Borescopes, fiberscope, or mirrors may be used to examine the welds of smaller diameter pipe. A television

camera on a travelling buggy may be beneficial for the internal inspection of critical large diameter pipe.

4. Check finished pipe for misaligned welds, excess weld reinforcement, undercut, cracks, blowholes, and other injurious defects.
5. Run-in and run-out tabs are generally utilized for submerged-arc welded pipe. Ensure that they are in place and properly aligned before welding begins. Also, confirm that they are of the proper material, i.e., the same as the plate to be welded.
6. Ensure that any drawing operation is carried out to the specification and that temperatures are as outlined.

Threaded Pipe (Pipe Casing and Tubing). In addition to the general inspection requirements, inspection of the threads is important. A list of items that should not be overlooked in thread inspection follows. In general, mill inspectors examine all threads on each length of pipe immediately after machining. Purchaser inspectors usually verify thread requirements on a sampling basis, say on five to ten percent of the order, selected at random.

1. Ensure that the proper thread has been produced.
2. Check the lead, taper, height, length, root diameter, form, etc., of the threads using thread gauges.
3. Verify the thread size with a ring gauge or calibrated coupling. Calibration is performed to a recognized standard and the coupling is identified by stamping or some other permanent form of marking. If possible, witness the calibration of the coupling.
4. Check for wavy threads. Wavy threads vary in height on the vertical or at the periphery. Variation may also occur on the longitudinal plane. An optical comparator may be required to detect problems. Careful measurement or a photograph of the threads has also been used to detect these problems.
5. Examine the threads to detect if any are cracked or broken. Magnetic particle or liquid penetrant testing may be used.
6. Inspect threads for flat spots, excessive scaling, etc. Require that all scale, dirt, etc., be removed to permit examination of the underlying surfaces. If turnings from machining are found in the threads, require that all of the threaded areas be cleaned.
7. Ensure that the proper thread protection, e.g., grease, thread covers, etc., is applied for storage and shipping.

Fittings and Flanges. In general, fittings and flanges are inspected as outlined in Pre-order Activities, Chapter 3. It is essential to ensure that pipe fittings will correspond to the mating surfaces of the pipe in dimensions, weld preparation, ovality, etc.

1. Examine fittings to ensure that they are the correct type (forged, rolled, welded, cast, etc.)
2. Ensure that the fittings bear the correct identification, which is generally rolled, stamped, stencilled, or affixed by some other suitable means. Identification usually includes:
   a) the manufacturer's name or trademark;
   b) schedule number or nominal wall thickness;
   c) size;
   d) grade; and
   e) heat number.
3. Verify the dimensions of fitting by measuring:
   a) the ID (at both ends);
   b) OD (at both ends);
   c) thickness; and
   d) ovality.

4. Verify that the fitting is the correct type, i.e., 90° ell, 45° ell, reducer, etc.

5. Examine the welds on fabricated mitred fittings for defects and ensure that the correct shape has been produced.

6. Check mitred construction for deformed areas, flat spots, raised areas, etc., that may indicate defects.

7. Verify the fitting end finish. Check the squareness of plain ends. For bevelled ends check the bevel angle, land area, etc., as applicable.

8. Examine the fittings for laps, laminations, gouges, dents, and other "injurious" defects. Defects that do not exceed 12.5% of the wall or penetrate the minimum wall thickness may be ground out. The grinding is feathered to produce a smooth, even contour. Larger defects may be weld-repaired provided that they do not penetrate more than one-third of the wall thickness or are more than one-fourth of the nominal diameter in length.

NOTE: For castings, an injurious defect is considered to be 20% of the actual wall thickness or 1 in. (25 mm), whichever is smaller. It is also considered an injurious defect when the wall thickness exceeds 10 in.^{2} (65 cm^2).

9. Ensure that fittings are properly bundled, crated, or otherwise packed for shipment. End caps may be required to protect bevels.

Fabrication of Spool Pieces (Welded)

Pipe spool pieces are generally fabricated by welding. Inspection involves verification of the welding, as per Welding Inspection, Chapter 8, and ensuring that the spool pieces are of the correct size. The following should provide additional guidelines.

1. Confirm that the correct pipe diameter, flanges, and fittings are being used. Examine mill test certificates and confirm that the material is properly identified.

2. Examine the weld procedures to ensure that they are correct and adequate for the job.

3. Measure the length of various spool pieces, allowing for bends or other fittings, making certain that they will produce products of the specified dimensions.

4. Examine weld preparations to ensure proper bevel angles, land areas, etc.

5. Examine the fit-up to ensure that the pipe sections, fittings, etc., are properly aligned before welding begins.

6. Examine tack welding, dogging clips, and other alignment devices.

7. Witness the root pass on some welding. The root pass on large diameter pipe will probably be back-gouged and welded from the inside.

8. Ensure that welds are free of cracks, undercut, blowholes, etc.

9. Ensure that weld reinforcement is not excessive.

10. Confirm that each welder has identified joints by stamping or some other suitable means.

11. Verify angular displacements with protractors or compasses, and compare them with isometric sketches.

12. Witness any hydrostatic tests on various spool pieces.

13. Verify radiography, ultrasonics, or other forms of NDT. Witness some of the testing.

14. Ensure that spool pieces have not been distorted and that any ovality is within acceptable limits.

15. Verify that bevel ends, flanges, or other facings

are properly protected for shipment and storage of the spool pieces.

16. For flanged spool pieces, confirm that the proper gasketing has been used.

17. Ensure that proper bolting of the correct size has been used.

18. Make certain that all painting, cleaning, or other operations have been performed satisfactorily.

NOTE: Some piping requires that mill scale be removed from the inside surfaces by sand blasting.

Spool Pieces Fabricated by Bending

Bending to produce "elbows," "U-bends," and even "circle-bends" has been used for many years with tubing found in heat exchangers, boilers, and other components because it is able to produce both simple and complicated shapes relatively economically when compared with welding. For thicker walls, pipe bending may be somewhat more difficult but it is increasing in popularity as it is readily performed both in the shop and the field.

Cold bending of pipe is performed routinely on material up to about 20 in. (50 cm) in diameter, while induction machines may hot bend material up to 70 in. (178 cm) in diameter. In the field, many companies limit cold bending to a maximum of 12 in. (30 cm) diameter material.

Definitions. In order to properly inspect bent pipe, it is necessary for the inspector to understand some of the terminology.

Degree of Bend: The degree of bend is the nominal bend angle such as 45°, 90°, etc.

Extrados: Extrados is the outside arc of the bend.

Intrados: Intrados is the inside arc of the bend.

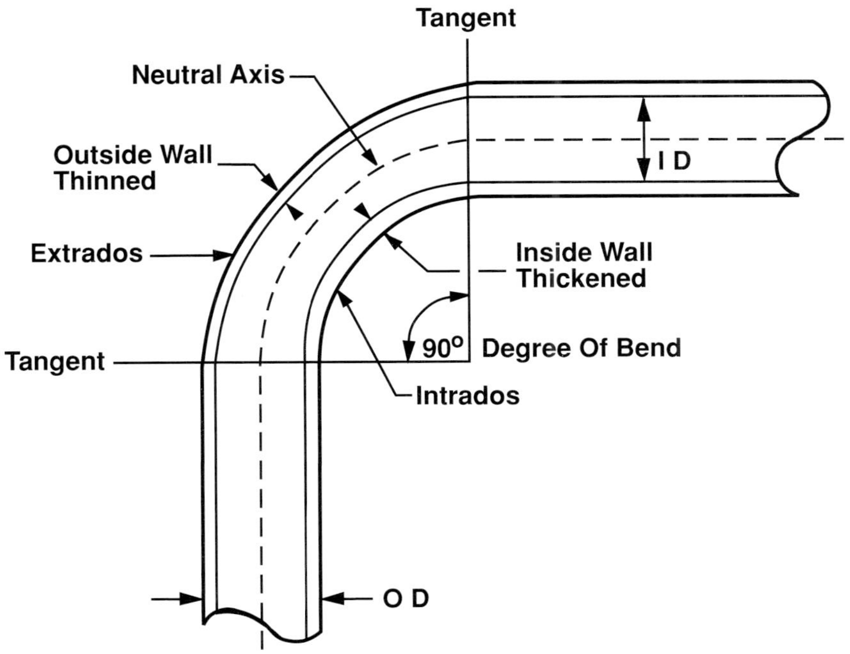

ELEMENTS OF A BEND

Neutral Axis: Neutral axis is the centerline axis of the pipe or tube along the arc of the bend.

Plane of the Bend: Plane of the bend is the straight sections' relation to the axis preceding the bend.

Radius: Radius is the arc of the bend itself. It is usually measured at the centerline or neutral axis of the material. The radius is often referred to in terms of multiples of the OD, particularly for tubing. For example, a 3D (3 times diameter) bend in 2-in. (50 -mm) OD tubing would be a 6-in. (15-cm) centerline radius. However, dealing with pipe to NPS can be misleading because the actual OD of NPS material is greater than nominal. For example, a 4-in. (10-cm) pipe with an OD of 4½ in. (11 cm) and a 3D bend in 4-in. (10-cm) pipe formed on a 12-in. (30-cm) radius, would have a true D factor less than 3D; in fact, it would be 1.33D. Therefore, it is important for the inspector to determine what parameters are used to establish the D factor.

Springback: Springback describes the attempt by the pipe to return to its original shape.

Tangent: The tangent is the straight material at each end of the bend. A bend with no straight portion at either end would have zero tangents.

Wall Thinning: Wall thinning is usually expressed in the form of a ratio: the amount of reduction from the nominal versus the amount of wall thickness remaining in the extrados of the bend.

Wall Thinning. Wall thinning is perhaps the most common complaint with respect to pipe bending. During bending, the extrados is stretched and therefore thins, while the intrados is compressed and therefore thickens. The neutral axis or centerline generally remains unchanged.

Expected thinning is usually determined by the following ratio:

$$\frac{\textbf{Outside Radius of Bend} - \textbf{Centerline Radius of Bend}}{\textbf{Outside Radius of Bend}}$$

Wall thinning may be greater than predicted because of such things as lubrication, tool drag, movement of the neutral axis, etc.

Wall thinning may also cause problems with respect to tolerances allowed for in various piping standards. For instance, ASTM A 53[11] and A 106[12] for carbon steel pipe all have virtually the same minimum thickness tolerance of 12.5 percent. If a bend is made in an area with this minimum thickness, then the extrados of the bend will thin out producing a thickness less than that required by the specification, resulting in rejection of the component.

Inspection of Bent Spool Pieces. Inspection of spool pieces formed by bending generally involves verification that the component has the desired shape, angle, dimensions, and is free of cracks or other injurious defects. The inspector is referred to standard ES-24,[13] of the Pipe Fabrication Institute (PFI)[(6)] which, although not mandatory, is generally considered an industry standard.

1. Examine the first, middle, and last pieces of any run, as machine settings may change.
2. Verify that the bends have been made to the correct angle such as 45°, 90°, etc. The bend angle usually has a tolerance of ± 1° because of springback. Compasses, protractors, or templates may be used to verify the angle.
3. Verify that the correct radius bend has been made using a compass or other accurate measuring device.
4. Check the ovality of the pipe using calipers and a steel ruler or, if the shop is so equipped, use a highly accurate theodolite (survey instrument) connected to a micro-processor.

[(6)]Pipe Fabrication Institute, Springdale, PA.

5. Check the configuration of complicated shapes ensuring that they comply with isometric sketches or other outlines of the component.

6. Look at the thickness of the components. Ultrasonics are the most effective way of detecting excessive thinning on the extrados or excessive thickening on the intrados. It may be difficult to check the intrados on small-diameter piping and short radius bends. Radiography may be used to verify wall thickness and quality on critical items. Thickness generally must meet the requirements of ANSI/ASME B31.1,[4] 31.3,[5] or other similar codes and standards. In general, the minimum thickness must equal the minimum required pressure design thickness, plus the sum of all allowances for mechanical, corrosion, erosion, or other considerations:

$$\mathbf{t_m = t + c}$$

where: t_m = minimum thickness
t = pressure design thickness
c = sum of all mechanical, corrosion, erosion, etc., thickness allowances

7. Check for flattened bends.

8. Using liquid penetrant, magnetic particle, or some other suitable process, check for cracks, tears, splits, slivers, and other anomalies that will affect the integrity of the bend.

9. Examine the components to ensure that any required cleaning, both internal and external, has been performed to remove tools, lubricants, packing material, scale, or other products.

10. Ensure that the component ends are square cut, beveled, etc., as required.

11. Check for wrinkling or waves in the bent material. Once a wrinkle has formed it cannot be removed during the bending process and is virtually impossible to correct afterwards.

NOTE: There are numerous company specifications governing measurement of wrinkles, many of which are inconsistent with thickness tolerances permitted on the pipe or tube. However, PFI ES-24[13] outlines general criteria for acceptance/rejection of wrinkles or waves. These criteria are as follows:

a) Wave shapes should blend into the pipe surface smoothly.

b) The maximum vertical height of any wave measured from the average height of two adjoining crests to the valley shall not exceed three percent of the nominal pipe size.

c) The minimum ratio of the distance between crests, as compared to the height between crests and the included valley, shall be 12-to-1.

12. Check the overall dimensions of the spool piece to ensure that it will fit where it is supposed to.

Flanges. Flanging for pipe bending generally involves flaring, especially to produce Van Stone flanges. This facilitates field installation as the slip-on flanges may be rotated to align the bolt holes for assembly.

Van Stone flanges are manufactured in the field by bending or flairing the pipe end. In austenitic stainless steels and other ductile materials, this has been quite successful. However, in carbon steels such as those of ASTM A 53,[11] A 106,[12] API 5L,[8] etc., bending or flairing often resulted in cracking. Even with an intermediate stress relief, cracking frequently occurred. The normal method of manufacturing these alloys was to fabricate the Van Stone flange and weld it to the pipe.

With the development of ASTM A 587[14] material, it became possible to produce the severe 90° bend to cold form the lap required for the flange, without cracking. Consequently, A 587 is used almost exclusively in the field to cold bend flanges. However, it is restricted in size to 4 to 6 in. (10 to 15 cm), although there is a move towards producing larger sizes.

With a hot process, lap joint flanges up to 18 in. (46 cm) may be made, but the process for large diameter flanges is relatively expensive.

Inspection of cold bent or flaired flanges usually involves the following:

1. Examination of the lap to ensure that no cracking, splitting, tearing, fissuring, or other defects have occurred. Liquid penetrant, magnetic particle, radiography, ultrasonics, or some other suitable method is used to identify such defects.
2. Hardness tests are often performed on carbon steels and austenitic stainless steels that are to be used in environments that may promote cracking such as sulfide stress cracking.
3. Check the dimensions to ensure that an adequate bearing surface is provided for the slip-on flange.
4. Check the lap material thickness to ensure that it is adequate for the stresses that will be applied to the joint.
5. If the lap is stress-relieved, make certain that a thermocouple is attached to the material to ensure an accurate recording of the temperature. Ensure that the correct procedure for the material is followed.
6. Ensure that the lap material has been cleaned of scale, grease, dirt, and other debris before the slip-on flange is aligned and the bolts tightened.

Pipe Tests

In order to assist the inspector in assessing test results, the following brief description provides an outline of mechanical testing procedures commonly applied to piping.

Flattening Test. A flattening test is performed by placing a section of pipe between two parallel flat plates and flattening it by moving the plates together. The test is normally done in two stages. The first, a ductility test, involves flattening the pipe until the distance between the plates is less than the value of H:

$$\mathbf{H} = \frac{(1 + e)\,t}{(e + t/D)}$$

where: **H** = distance between the flattening plates in inches (or mm),

t = specified wall thicknes in inches (or mm),

D = specified or calculated outside diameter in inches (or mm),

e = deformation per unit length and is constant for any given grade of steel, i.e., 0.09 for low carbon (0.18 max. carbon), 0.07 for medium carbon (0.19 min. carbon), 0.08 for ferritic alloys, and 0.09 for austenitic alloys.

The second stage is a soundness test where one continues to move the plates together until the specimen breaks or the opposite walls of the pipe meet. No cracks or breaks are permitted on the inside, outside, or ends of the sample, unless they are superficial ruptures caused by surface imperfections. These are judged in accordance with finish requirements.

If the D-to-t ratio is less than 10, the strains imposed at the 6 and 12 o'clock positions on the inside surface of the pipe are inordinately high because of the pipe geometry. Therefore, such values would not be cause for rejection.

Test specimens may vary somewhat in size. The recommended section is a minimum of 2½ in. (6.3 cm) long for seamless pipe and a minimum of 4 in. (10.2 cm) long for welded pipe.

For the latter, the weld is placed at 90° to the direction of the applied force.

Hydrostatic Test. If a hydrostatic test is required, each length of pipe is tested by the manufacturer at a test pressure that induces a minimum wall stress in the pipe of 60 percent of the minimum specified yield strength for carbon steel and ferritic alloy materials, or 50 percent of the specified minimum yield for austenitic alloys. The test pressure is determined by:

$$P = 2\,St/D$$

where: **P** = hydrostatic test pressure in psi (or MPa),
S = pipe wall stress in psi (or MPa),
t = specified nominal wall thickness in inches (or mm),
D = specified outside diameter or 2t plus the inside diameter.

The test pressure is generally rounded to the nearest 50 psi (0.3 MPa) for pressures below 1,000 psi (6.9 MPa) and to the nearest 100 psi (0.7 MPa) for pressures above 1,000 psi. Test pressures are generally held for a minimum of five seconds. The test pressure on welded pipe is held as long as necessary for the inspector to examine the entire weld seam.

Tensile Test. A tensile test is performed by placing a reduced-section specimen in a test machine that pulls the material to failure. The yield point of the material is determined by the drop of the beam, the halt of the pointer or gauge, or by any other approved method. If the material does not display a definite yield point, the yield strength is usually defined as that corresponding to a 0.2% permanent offset of the specimen gauge length or to a total extension of 0.5% of the gauge length under load.

A tensile test is used to determine the yield strength (YS) of the material, the ultimate tensile strength (UTS), and some measure of ductility that may involve elongation or reduction in area. Elongation is determined by fitting the pieces of the fractured specimen together again and measuring the distance between the gauge marks. The elongation is expressed as a percentage of the original gauge length.

Reduction of area (RA) is determined by fitting the ends of the fractured specimen together and measuring the mean diameter or the width and thickness at the smallest cross section. The difference between the smallest cross section and the original cross section, expressed as a percentage, is the RA.

Additional Tests. Additional tests that may be specified to meet specific requirements include hardness tests, impact tests, and corrosion tests. Hardness tests are usually employed when the material will be used in hydrogen sulfide or similar service in which specific structures in the alloy may experience various forms of stress cracking. For instance, it has been determined that carbon and low alloy steels with hardness levels less than Rockwell hardness C scale (Rc) 22 or equivalent hardnesses on other scales are generally resistant to such cracking. Therefore, macrohardness tests may be required on piping.

A hardness test is usually performed on the pipe ends of seamless material. Readings are taken on both the inside and outside surfaces where practical. Hardness readings may also be taken from the ends of plain end pipe and from the bevels of beveled pipe.

Hardness readings for welded pipe are taken on the parent material, the weld, and the heat-affected zone (HAZ). All hardness readings are performed after heat treating, straightening, and other such operations.

Some purchasers prefer to have pipe test coupons processed along with the pipe requiring hardness tests since they are concerned that stresses produced by tests on the pipe could cause failure in service. Such coupons may be examined microscopically as well.

Hardness readings are usually based on an average of three recorded readings. The decision to reject hardness values over Rc 22 varies but, in general, one value of Rc 23 to 24 is permitted among the three. However, if the material shows consistent values above Rc 22, many companies will reject the lot, regardless of the average of the three readings. Other alloys with higher hardness values are permitted as outlined in NACE International (NACE) MR0175, "Standard Material Requirements: Sulfide Stress Cracking Resistant Metallic Materials for Oilfield Equipment."[15]

Impact tests for piping generally entail the Charpy or Izod tests, in which notched or grooved specimens are struck a single blow with a hammer in a testing machine and the energy absorbed in breaking the material is then determined. Testing temperatures are usually below ambient, but the testing temperature and anticipated service temperature need not be the same.

Acceptance is based on the average of three specimens, although not more than one specimen may exhibit a test value below the specified minimum. If an individual value falls below either two-thirds of the specified minimum or 5 ft-lbs (6.8 J), the test is rejected. Retesting is permitted, but if the same or similar results are obtained, the lot is rejected.

Austenitic stainless steels and other high alloys are often tested to determine their susceptibility to corrosion by standard solutions. For example, ASTM A 262[2] tests the susceptibility of alloys to intergranular corrosion. Other standards or recommended practices are available for the evaluation of various products; in particular, NACE International has issued a number of recommended practices detailing corrosion testing of materials.

Corrosion testing may be applied to samples cut from lengths of pipe or it may be applied to coupons made and treated along with the pipe.

CHAPTER 8

Welding Inspection

Overview

Welding inspection is critical during shop fabrication of pressure vessels and piping. In fact, aside from dimensional aspects, the most common quality problems associated with fabrication relate to welding. This is easily understood when one considers the complexity of the welding process, which involves the application of large quantities of heat to a relatively small area, resulting in melting or fusing of metal and its subsequent resolidification.

Many variables affect the success of the welding operation. These variables can be controlled by the welding process, the welder, and/or the procedure.

Process Variables

The following process variables are important in welding inspection.

1. Type of current (AC or DC).
2. Type of process: Shielded Metal Arc Welding (SMAW), Submerged Arc Welding (SAW), Gas Metal Arc Welding (GMAW), Gas Tungsten Arc Welding (GTAW), Electron Beam (EB), Friction, Electroslag, Spot, Stud, etc. The most common types of welding used in the construction of pressure equipment and piping are SMAW, SAW, GMAW, and GTAW.
3. Polarity of the electrode (positive or negative), which is a procedure variable.
4. Composition of the flux or shielding and backing gases, which is also a procedure variable.

Welder Controllable Variables

The welder is responsible for the following:

1. dimensions of the weld joint;
2. cleanliness of the surfaces to be welded;
3. travel speed;
4. surface quality of the deposit, i.e., freedom from cracks, blow holes, porosity, or undercut;
5. sheltering work from contamination; and
6. use of appropriate consumables.

Procedure Controlled Variables

Procedure controlled variables include:

1. base material;
2. filler metal;
3. welding current and voltage;
4. type of current (AC or DC);
5. travel speed;
6. shielding gas, flux, etc.;
7. type of backing, i.e., shielding gas, backing bar, consumable inserts;

8. type of joint, e.g., single Vee, double Vee, J, and U;
9. preheat, postheat, or both;
10. type of deposit, i.e., stringer bead, weave, or a combination;
11. welding position, i.e., flat, vertical, overhead and horizontal, or any combination thereof; and
12. type of equipment, i.e., manual, semiautomatic, or fully automatic.

Welding Procedure

There are many variables to be controlled during welding, and it is important to produce welds with consistent properties from job to job and welder to welder. This can be accomplished largely through the use of written welding procedures.

A welding procedure is a step-by-step plan that welders follow to ensure the integrity of the finished product, i.e., the weld. Using an established procedure, any competent welder should produce welds with essentially the same properties as those produced by any other competent welder using the same procedure. Therefore, a basic feature of weld inspection is to determine that each and every welding operation will be performed to an established and approved welding procedure.

An approved welding procedure consists of two components—the Welding Procedure Specification (WPS) and the Welding Procedure Qualification Record (PQR)— (Appendix 5). These documents outline the welding sequence and what procedural variables produce satisfactory test results. The essential elements that are documented to produce an acceptable procedure include the following:

1. Procedure number and company name.
2. Title: Includes the procedure number, the material specified, and the thickness qualified.
3. Scope: Outlines the base material, thickness qualified, company performing the welding, and the governing regulations.
4. Category of the base material, e.g., ASME Section IX, Group P1 to Group P61 (Appendix 6).
5. Filler metal, both by trade name and ASME designation, if applicable, including the weld metal analysis, e.g., an E7O18 material might be identified as: E7O18 and manufacturer's trade name to ASME SFA 5.1, Electrode Classification F4, Weld Metal Analysis A-1.
6. Welding position: Flat, vertical, overhead, horizontal, or all positions.
7. Welding process: SMAW, SAW, GMAW, etc.
8. Shielding gas composition and flow, flux composition.
9. Use of backing strips, backing gas, etc., and flow rates, as applicable.
10. Dimensions of the welding groove.
11. Electrical characteristic: AC or DC current, current and voltage for various electrode diameters, wire feed rates, and polarity of the electrode.
12. Preheat and interpass temperature.
13. Postweld heat treatment.
14. Arc travel speed, component travel speed, etc.
15. Any special requirements such as stringer bead or weave technique, cleaning of each pass, final appearance of the weld deposit, and any special testing required.

Welder Qualifications

Every welder working on pressure vessels or pressure piping must be qualified through testing for each process and welding procedure that will be used on a project. Testing generally includes the production of a test weld following an established welding procedure and the removal of coupons to permit mechanical testing (usually bend testing). Radiographic examination may be used in place of mechanical testing. Specialty tests such as corrosion testing, impact testing, or other tests to determine specific properties of a weld joint that could affect its service life may also be necessary.

Requirements for performance qualifications are outlined in Section IX of the ASME Code.[16] In general, an inspection will verify the following:

1. Qualifications of all welders who will be used on the project by examining work in progress, plant records, certified test results, etc.;
2. Symbols to be used by each welder to identify welds;
3. Joint positions for which a welder is qualified, i.e., flat, vertical, horizontal, overhead, or all positions;
4. The type of equipment a welder is qualified to use—manual, semi-automatic, or fully automatic;
5. The type of test used to qualify a welder, e.g., bend test, radiography, etc.; and
6. The results of any specialty tests used to qualify welders as outlined in contract specifications.

Welding Inspection

Following verification of the welding procedure, weld inspection generally takes place.

Joint preparation. In preparation for welding, check that:

1. The joint edge-preparation configuration conforms to the drawing and to code requirements. This may be done with any accurate protractor or instrument such as a profile gauge. See Appendix 5.
2. The joint edge and the area within 1 in. (25 mm) of the edge on both sides of the material are clean and free of grease, paint, oil, scale, rust, and similar foreign matter (3 in. [75 mm] may be required for some high alloy material).
3. The joint edges prepared by thermal cutting are ground to bright metal for carbon steel welds, or are ground 1/16 in. (1.5 mm) past the deepest indentation using iron-free wheels for stainless steels and other high alloys.
4. The alignment of the joint is in accordance with the design code and the applicable welding procedures.
5. All material to be welded is properly identified and traceable to the material examined during the inspection of the raw material.

On critical components, it is often advisable to perform some form of NDE such as magnetic particle testing or liquid penetrant to ensure that the weld preparations are free from laminations or other potential defects. This is particularly important for components that could experience hydrogen damage in service. These tests may be witnessed on a sampling basis or completely at the inspector's discretion.

Prior to Welding. Prior to welding, ensure that:

1. Only certified welders will perform the work.
2. The appropriate, approved welding procedure is present at the work station.
3. The metal temperature has reached ambient temperature before welding begins. Metal brought into a warm shop from outdoor storage, particularly in cold climates during the winter, will take considerable time to reach the shop floor temperature. If condensation occurs during the warming process it can cause

underbead cracking during welding. Similarly, metal brought in wet from rain should be allowed to dry before welding starts. It may be necessary to apply heat to dry components.

4. The correct welding electrodes, wire, flux, shielding gas, and other consumables, as applicable, are available. On critical components, mill test certificates, gas analysis sheets, flux composition, etc., are generally available for review.
5. Preheating equipment, if required, is at the work station and is operable.
6. Proper welding shields have been installed, as necessary, to prevent contamination.
7. Lighting is adequate.
8. Run-in and run-out tabs, if required, are in place and are the same material as the base metal.

During Welding. During welding, be certain that the following are accounted for.

1. Tack welds for fit-up are kept as small as possible.
2. All tack welding is performed by welders qualified to the same procedures as are to be used for the finished weld. However, tack welds may also be made using SMAW, GTAW, or GMAW processes, but the joint weld must be completed with submerged arc, electroslag, or some other shielded process. In these cases, tack welds, performed to an approved procedure, are then incorporated into the finished weld.
3. Tack welds are not cracked or otherwise defective.
4. Tack welds are either removed completely before welding or they are thoroughly fused and incorporated into the finished weld, as applicable. For instance, on vertical welds with SMAW, tack welds are usually removed to prevent cracking.
5. Poison pads are used when attaching carbon steel or other low alloy fittings or welding aids to stainless steel and other high alloys.
6. All welding consumables are properly protected from moisture and other contaminants.
7. Low-hydrogen welding consumables are stored in ovens at manufacturer's recommended temperatures (usually 250°F [121°C]).
8. Low-hydrogen welding consumables in use are held in heated containers or are returned after four hours to a baking oven for treatment at the manufacturer's recommended temperature.
9. Only enough low-hydrogen consumables are issued for a four hour period. Other welding consumables are issued in quantities sufficient only for a one shift operation. This helps to avoid the use of incorrect materials. Welding consumables are returned to stores and properly protected at the end of each shift.
10. Electrodes with damaged, wet, or oily coatings are not used and are discarded.

NOTE: A shop with welding consumables lying around tends to produce poor quality. A welder in a hurry to finish may pick up and use discarded material rather than go to the stores for properly stored components.

11. Preheat and interpass temperatures, as applicable, are maintained and monitored with thermocouples, contact thermometers, infrared scanners, heat sensitive crayons, or some other suitable means.
12. Welding performed outdoors is properly shielded from the elements.
13. Welding is not performed if the ambient temperature is below 0°F (–18°C) in the field or in an unheated shop.
14. Preheating to a minimum of 60°F to 100°F (16°C to 38°C) is performed if the ambient temperature is between 0°F and 32°F (–18°C to 0°C) in the field or an unheated shop.

15. All slag is removed from each pass by chipping or brushing (grinding to remove slag is not recommended, as it may lead to grinding cracks, undercut, or some other linear defect).

16. All porosity, blowholes, gas pockets, etc., are removed from each pass before the next pass is made.

17. There is no undercut along the edges of each pass.

18. The welding sequence is observed to prevent distortion.

19. There is no grinding, painting, or other process going on close enough to the welding to result in contamination.

20. The welding is sheltered so that the opening of shop doors, passing vehicles, etc., cannot create sufficient draft to disrupt shielding gas or to cool welds enough to create hard spots, high stresses, or other detrimental conditions.

21. Automatic welding is under the control of an operator at all times to prevent or detect arc wander, component movement, or other problems.

22. Any special requirements or restrictions imposed by the specification and/or the applicable code are observed.

Finished Welds. On finished welds, check that:

1. The as-finished dimensions conform to the approved drawing.

2. Joint distortion from welding does not exceed the allowance.

3. All required NDT has been performed and is acceptable. All test reports are signed by individuals certified under the appropriate ASNT,[7] CGSB[8], or other acceptable qualification requirements.

4. All welders have identified their welds by stamping or placing a tag adjacent to the weld. On some specialty items, identification may take the form of providing drawings identifying the welds and welders rather than affixing something on the component. This is usually the practice for items constructed of extremely thin material or for items that might experience cracking from extraneous stress conditions caused by stamping or additional welding.

5. All arc strikes have been blended into either the weld metal or the parent material.

6. All welding aids such as dogging clips, spacer bars, and support beams have been removed and attachment welds blended into the metal.

7. Postweld heat treatment, if required, has been performed and is acceptable. Heat treat charts are available.

NOTE: Many fabricators prefer to perform all NDE before heat treatment so that repairs can be made and then the heat treatment performed, but cracking can and does occur during heat treatment. It is always good practice to perform the final NDT after heat treatment. Finally, the hydrostatic test is performed after heat treatment.

8. All required tests have been witnessed by the authorized inspector.

9. All required signatures have been placed on route cards before the piece is transferred to the next operation.

10. Weld surfaces on both the inside and outside of components have been inspected, wherever practical. Nozzle welds, in particular, have been finished on both the inside and the outside of the vessel. Weld profiles generally have a smooth, even appearance with a slightly convex contour (butt welds).

---

(7) American Society for Nondestructive Testing, Inc., Columbus, OH.

(8) Canadian General Standards Board, Ottawa, Ontario, Canada.

11. Joints between dissimilar thicknesses of materials have been blended to provide a smooth, even profile usually with a 3-to-1 or 4-to-1 taper.

12. Measure fillet weld legs to ensure that they conform to specification. Normal practice is to have the fillet size equivalent to the plate thickness. For dissimilar thicknesses, the weld may be equal to the thickness of the thinnest plate, as a minimum.

13. Examine some of the welds with a hand lens for signs of cracking, particularly after the hydrostatic test.

14. In multiple bead welds, particularly if a series of stringer beads is used, ensure that there is no groove or depression between adjacent beads. Grinding and deposition of a cover bead may be required.

15. Ensure that the reinforcement on the weld is within specification; $^1/_{16}$ to $^1/_8$ in. (1.5 to 3 mm) maximum, is generally the acceptable amount depending on base metal thickness.

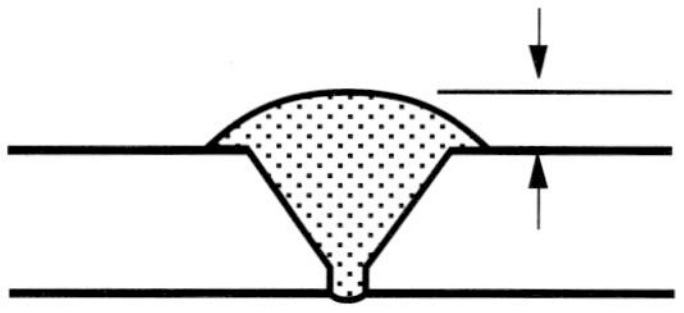

WELD REINFORCEMENT

16. Ensure that the weld cap extension over the sides of the weld groove is within specification. Again, $^1/_{16}$ to $^1/_8$ in. (1.5 to 3 mm) is the normal maximum, although some companies prefer to have the extension ground away in order to check the weld edge for cracking.

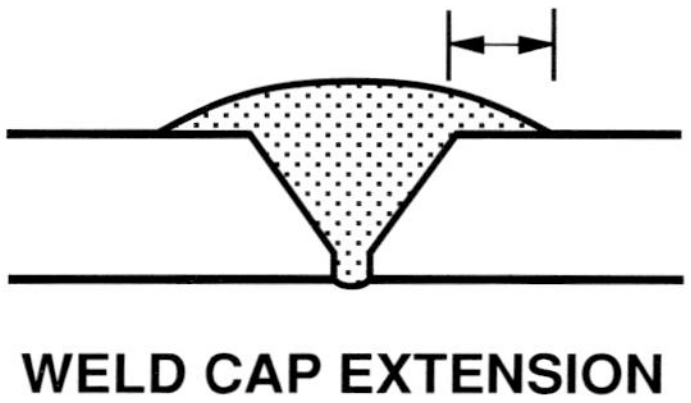

WELD CAP EXTENSION

17. Examine the weld beads for signs of peening. Peening is usually not acceptable, except as outlined in certain special applications and specifications.

18. Examine the weld beads for signs of excessive grinding. Some companies require complete removal of the weld cap to produce a smooth blended surface with the parent material. Any decrease in material thickness at the weld joint is cause for rejection. Ultrasonics may be used to verify the thickness of smoothly ground joints.

Tips to Help the Welding Inspector

The following points will assist the welding inspector in identifying problems or conditions that may lead to problems. Also, some general practices are given to prevent or correct problems.

1. Double welded butt joints are back-gouged or gouged to a clean surface before the second side is welded to ensure proper root penetration and cleanliness.

2. If welding is stopped, care is necessary in restarting to obtain the required penetration and fusion. For submerged arc, chipping out a groove in the weld crater before restarting is recommended.

3. Welder's identification is stamped at intervals not exceeding 3 ft (90 cm) on a vessel. Stamping is practiced only on materials over $^1/_4$ in. (6 mm) thick. Even then, low stress stamps are recommended.

4. Hand-fed filler metal is cleaned, frequently by wiping with a solvent-soaked clean cloth before welding, to prevent weld contamination.

5. Welding wire may be fed through a cleaning device during welding. The tip of the wire is removed before restarting an arc.

6. Push-pull wire feeders often produce the better

quality welds when welding austenitic stainless steel and other high alloys. A push system is most successful with aluminum.

7. Shielding gas may contain significant quantities of moisture, resulting in weld contamination. Filters and dryers may be required downstream of the regulator, particularly for critical items.

8. Shielding gas is generally located as close to the work as possible to prevent condensation and dirt accumulation. Central gas distribution lines are often the source of contamination.

9. Grinding to remove slag often results in grinding cracks, undercut, and other unacceptable defects.

10. Zinc and aluminum from grinding or painting operations have caused cracking of welds, particularly in austenitic stainless steels. Molten copper will penetrate carbon steels, causing intergranular cracking. Sulfur contamination creates problems in welding high-nickel alloys. High-alloy materials are often fabricated in areas that can be isolated from the rest of the shop.

11. Some liquid penetrant test solutions have caused cracking and porous welds when not properly removed before more weld metal is deposited. Such solutions, when used on austenitic alloys, generally contain low quantities of halogens and sulfur to avoid defects.

12. Multiple repairs can produce excessive distortion, particularly with austenitic alloys. If more than two repairs are required in any location, special distortion control measures will probably be necessary.

13. Peening is often used to control stresses. For joints that will be in environments conducive to stress corrosion cracking, peening is not recommended. Thermal stress relief would be the most effective way of relieving stresses and even this cannot always prevent cracking.

14. When welding dissimilar metals, the high-alloy material is frequently "buttered" over the low-alloy material to avoid contamination, and then the weld is made to avoid weld contamination.

15. When welding clad material, the standard procedure recommended by many fabricators involves the removal of the high alloy material for a short distance on either side of the weld groove. The low-alloy material is welded. The high-alloy material is then buttered over the low alloy, and finally the high-alloy material is welded.

Welding Parameters. The following is a list of welding variables that should be given consideration.

1. The machined angle on a weld groove must be wide enough to allow the electrode to reach the root of the joint and permit fusion with the side walls. An excessive angle requires more filler metal which can lead to hot cracking.

2. Weld fit-up, often referred to as the root gap or opening, affects the quality of the weld. Too great a gap will result in burn through; too little, in lack of penetration. Root gap is often obtained by inserting a spacer and tacking the joint. During welding of a longitudinal joint, the root gap could open or close depending on shrinkage. Evidence of excessive or uneven opening or closing is an indication of potential problems.

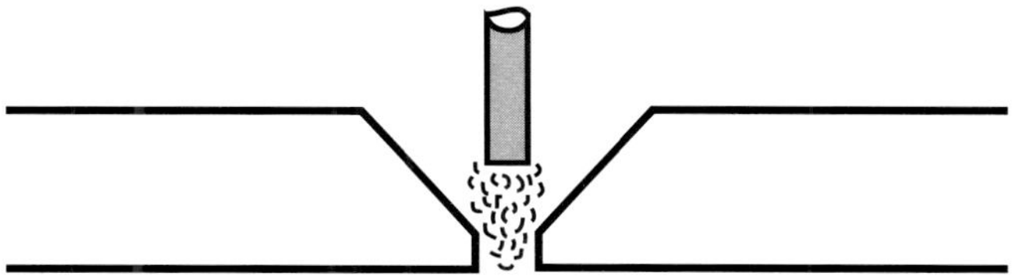

3. The root face or land area is the amount of parent metal left in the root when the joint is machined. It is fused when the root bead is deposited. Too little land area results in burn

through; too large a land results in incomplete penetration.

NO LAND AREA

EXCESSIVE LAND AREA

4. If travel speed is too slow, molten metal and slag may run ahead and start to cool. The weld puddle may run over this material with little or no penetration. Weld metal will tend to wash over the sides of the groove and may cause undercut as well as overlap. Too rapid a travel speed produces porous, non-penetrating welds.

5. When welding vertically, the crater is tilted to help slag run off.

6. Undercut is frequently caused by too long of an arc, or the welder tending to weld to one side of the groove. It is often corrected by reducing current, travel speed, electrode diameter, and electrode angle.

7. Weld spatter, while not a defect, may result in costly clean-up. Weld spatter is generally produced by too high a current, too long an arc, incorrect electrode angle, wet electrode, or incorrect polarity.

8. Rough welding is often the result of incorrect polarity and current. Wet electrodes and excessive travel speed also produce rough welds.

9. Arc blow occurs with DC welding when stray magnetic fields cause the arc to wander. It is often controlled by wrapping the welding cable around the component so that the magnetic field generated in the cable counteracts the stray field. However, care must be used to wrap the cable in the correct direction or else its field will reinforce the one causing the arc blow. Other methods of controlling arc blow involve changing to AC welding, reducing the current and decreasing the electrode size, reducing the arc length, welding in the direction of the blow, and shifting the work connection to the opposite end of the component.

10. Porosity is generally caused by contamination, too rapid freezing of the puddle, or too long of an arc. It is corrected by cleaning the joint, cleaning the welding wire, and keeping the weld puddle molten longer to allow the escape of trapped gases.

11. Improper fusion is caused most often by welding at too low a heat, too rapid a travel speed, and an improper joint gap. It is corrected by increasing the current, changing to stringer beads to promote deeper penetration, and ensuring a better root gap fit-up, particularly in long joints.

12. Improper penetration is caused by welding at too low a heat, too rapid a travel speed, too small a root gap, and improper angle on the weld groove. It is corrected by using higher currents, reducing the travel speed, increasing the gap spacing and using smaller diameter electrodes.

13. Cracking is considered the most serious defect in welding as it may lead to catastrophic failure. Cracking occurs for many reasons, the most common of which are lack of hot ductility of the material leading to hot tears, shrinkage in the center of the weld, the presence of dissolved gases such as hydrogen, poor fit-up leading to high stresses, and a lack of skill by the welder. Cracking is most often controlled by preheating to reduce stresses and drive off gases, increasing the number of passes to ensure shallow penetration, use of low-hydrogen material, and welding on dry surfaces. Making suggestions concerning welder skill is a delicate matter. Suggestions to help prevent cracking by an unskilled operator may include: back stepping before breaking an arc to prevent crater cracks, fusing in all stops and starts, and extensive inspection before welding to ensure proper joint design and fit-up.

14. Distortion occurs in welds simply because small areas of components are heated, while the remainder stays cool so that some movement or permanent change of shape results. Distortion control usually involves such things as restraining the work to prevent movement, sequencing of welds, welding on one side and then the other, starting at the center of long welds and welding towards the free end, welding on one side while heating the other, welding at two or more locations simultaneously. The idea is to equalize the welding on either side of the neutral axis.

15. Even with all precautions and proper techniques, some straightening will probably be required. Straightening of carbon steel and low-alloy components is usually accomplished by restraining the ends of the component and heating it at various points. Each heat is called a shot. Hammering components into shape with or without heat is not considered good practice.

NOTE: ANSI/AWS(9) A2.4,[17] "Standard Symbols for Welding, Brazing and Nondestructive Examination," is a very useful resource, and it is recommended that inspectors have access to it.

(9)American Welding Society, Miami, FL.

CHAPTER 9

Nondestructive Testing (NDT)

Methods

Nondestructive testing is a most valuable aid in assessing the quality of raw material, welds, machining, and other aspects of fabrication. For the typical shop fabrication of pressure vessels and piping, NDT methods include radiography, ultrasonics, magnetic particle, and liquid penetrant.

In order to perform satisfactorily, the owner's inspector will, in all probability, be required to approve and accept test results from these four methods. The inspector does not have to be fully qualified in all test disciplines to assess nondestructive test results, however some knowledge of the methods and their applications are necessary. In addition, knowledge of potential problems that may arise will assist in appraising the test results.

The following discussion of the four common test methods is aimed at helping the inspector to identify potential problems. It is not intended to provide detailed inspection techniques or to provide advice to certified technicians on how to carry out their functions. In addition, it is recognized that other nondestructive techniques and methods are applicable to the fabrication of pressure equipment, but the four discussed here are those most commonly encountered.

Testing of Pressure Components

Radiographic Testing (RT). Radiography is an extremely valuable inspection tool used to ensure quality in pressure vessels and piping. The work is usually performed by the fabricator's own personnel or by contract inspection personnel. Whichever the case, it is important that those performing the work are versed in both the safety aspects and the requirements for producing film of sufficient quality so that any defects are readily visible. The purchaser's shop inspector need not be totally familiar with all aspects of radiography but being able to interpret film would certainly contribute to an inspector's authority.

Radiography is performed under license from federal, state, provincial, and even municipal governments and various agencies. Those performing the work are certified under the ASNT, CGSB, and other requirements.

NOTE: See Appendix 7 for ASNT and CGSB requirements.

Consequently, the first thing the shop inspector does is to confirm the qualifications of those performing the work and to ensure that the proposed procedure will produce film meeting contract requirements.

Particular attention is paid to coverage expected during radiography. Some specifications and procedures state that 100 percent radiography is required. This may mean different things to the radiographer and the inspector. For instance, some radiographic procedures, codes, and standards interpret this as applicable to butt welds only so that no inspection of fillet welds such as are found on nozzles takes place. Other specifications interpret 100 percent coverage to mean that all welds are radiographed but not all of each weld. Still others permit 100 percent coverage to mean that only 85 to 90 percent of the total weld area is inspected. Fillet welds are often very difficult to inspect completely by radiography because of access restrictions. Some combination of radiography and ultrasonics may be used to ensure full coverage of fillet welds. Consequently, it is important to examine the shooting technique to determine whether or not the coverage desired by the customer will indeed be provided. It is important to remember that the purpose of radiography is to establish the size, shape, and location of discontinuities.

While it is unlikely that the purchasing inspector will perform the actual radiography, involvement in selecting the welds to be radiographed for spot examination, approving the shooting techniques, and the review and acceptance of film are likely activities. Therefore, the following list outlines a number of steps that can be taken to help assure the quality of the component.

1. For spot radiography, ensure that at least one spot is examined on each longitudinal and circumferential joint. Codes require one spot every so many feet (generally 50 ft [15 m]) of weld. On small vessels, fabricators may attempt to examine only one spot and apply this to the entire vessel because it is within the footage limit for the one spot.
2. Select for testing areas of high stress, particularly T-junctions, areas with obvious surface problems such as stops and starts, craters, excessive reinforcement, etc. Vary the spot selection to prevent welders from becoming familiar with your choices.
3. Examine at least one spot for each welder working on the vessel.
4. Examine at least one spot for each welding technique used, i.e., SMAW, SAW, GMAW, GTAW, etc. If each welder uses more than one technique—spots for each welder using each technique are examined.
5. Examine the welds that are radiographed to determine whether or not any identifying marks or defects are visible. They could help permit detection of "good films" that are submitted repeatedly as substitutes for poor or rejected film. Also, note the shape and appearance of certain defects to see if they keep reappearing.
6. Check the density of the films to ensure that they are adequate for interpretation. Density minima will vary with the source and film technique but generally are 2.0 for single-film viewing with film produced by a gamma source, and 2.6 for composite film viewing. The maximum density is generally 3.8 to 4.0 for either single or composite viewing with a minimum density of 1.3 for any film of a composite set (ASME Code Section V, Article 2).[18]
7. Welds to be radiographed are generally free of all spatter, slag, undercut, or other irregularity that might affect the film interpretation.
8. Ensure that all films display a penetrameter to ASME Section V, Table T-233.1 requirements, with selection based on Table T-276.[18] Penetrameters are placed near the outer edge of the film area.
9. Check radiographers' qualifications. All radiographers are certified to the ASNT, CGSB, or similar requirements. Supervision of work, procedure preparation and film interpretation is normally performed by individuals qualified to level 2 or 3 of the standards.
10. Ensure that film identification and traceability are adequate. This normally involves:
    a) client's name;
    b) job number or some other form of identification;
    c) part or heat number;
    d) weld or view number;

e) date of inspection;

f) letter R to indicate a repair;

g) radiographer or radiographic company's identification; and

h) a marking scale so that any section of film may be referenced from a zero or starting point.

11. Ensure that radiography reports contain sufficient information to permit identification of defects such as:

    a) radiography company's name, address, and telephone number;

    b) report number;

    c) radiographer's name and qualifications;

    d) type of source and strength;

    e) type of film used;

    f) radiograph number;

    g) description of defects and notation as to whether they are acceptable or rejectable;

    h) technique used—single wall, double wall, panoramic, etc. with a sketch to show the shot geometry; and

    i) code to which film is interpreted.

12. Film is normally interpreted to the requirements of ASME Section VIII, Division I, Paragraph UW-5I or DW-52[19] for vessels and to ANSI B31.1,[4] 31.3,[5] etc., as applicable for piping.

13. The film is placed as close as possible and generally not further than 1 in. (25 mm) from the back side of the component.

14. The maximum area covered by a single film is generally limited to points where the radiation path is within 6 to 12 percent of the wall thickness at the center of the film. This is governed by the curvature of the vessel and the distance the source is from the film. In general, two shots are required for small diameter pipe, for example, but increases to three shots for diameters greater than about 4 in. (10 cm) because of the 12 percent consideration, and to four shots for pipe over 8 in. (20 cm) in diameter, etc. This leads to some consideration in the direction of shooting to best detect flaws:

    a) Centerline cracks, incomplete fusion, incomplete penetration, etc., are best seen with a view normal to the weld surface.

    b) Slag, gross porosity, and gross inclusions are generally evident no matter what the view.

    c) Weld edge cracks, lack of edge fusion, etc., are best seen with a view parallel to the joint face although this is usually impractical.

    d) Square butt joints, O-groove, and J-groove joints usually can be inspected with a normal view.

    e) Other joints usually require two views, one parallel to each original prepared edge, although normal practice is to use one view.

    f) For fillet welds the best view is obtained if the viewing direction bisects the angle between the legs of the section.

    g) Fillet T-joints welded on both sides may require two shots for adequate coverage.

    h) Complete coverage of fillet welds may be impossible because of access.

15. Film is interpreted by examining it with a proper

viewer. Holding film up to plant lights or windows is inadequate.

16. Film identification that obscures the weld is cause for rejection of the film.

17. All defects are marked or identified on the film and are marked on the welds themselves. If disputes arise over film interpretation, request reference radiographs for comparison.

18. If possible, some defect removal is witnessed in the shop.

19. The radiographs of all repair welds are reviewed.

20. Defect maps, if required, are prepared showing the size and location of all defects, including those that were repaired.

21. Ensure radiographed welds are identified with the appropriate symbol (usually "R"—radiographed, or "T"—tested, although "R" may indicate repaired).

22. Following completion of the radiography, ensure that all films have been examined and represent the correct number for the contract. Ensure reader sheets are included with each film set. Ensure that film is stored properly or is shipped to the purchaser before signing any release form.

Ultrasonic Testing (UT). Ultrasonic testing as applied to shop fabrication of new pressure components, usually involves flaw detection through the use of a high frequency sound beam directed into the test material on a predictable path. If a flaw is encountered, the continuity of the sound in the material is interrupted and the back reflections from the discontinuity will appear on the screen at a different location, different height, etc. (depending on the system used). Interpretation of the screen display depends greatly on the skill and training of the operator. Operators are generally qualified to ASNT, CGSB, or some other equivalent requirement. Many companies have their own in-house training programs for UT operators and formal qualifications from recognized agencies or societies may not be obtained. This does not mean that the operators are less skilled than formally certified personnel. It does mean, however, that the purchasing inspector may have to assess the operator skill to ensure compliance with purchase specifications. This is usually accomplished by asking for a demonstration in which the operator uses the procedure to be employed to identify defects known to the inspector in sample pieces.

Again, the purchasing inspector does not have to be a skilled operator to assess the quality of ultrasonic testing and flaw detection, but a basic knowledge of the process and some familiarity with the potential problem areas are required. There are a number of factors that may be monitored to ensure that the correct procedures and techniques are being followed. See below.

1. Examine the test procedure to determine its adequacy.

2. Most flaw detection is performed with transducers of about 2.25 MHz. This frequency is generally adequate for both longitudinal and shear wave testing modes. Some miniature search units require higher frequency transducers to reduce sound beam divergence.

3. For flaw detection on welds, three basic search angles are used, 70°, 60°, and 45°, as measured from a line at right angles to the material surface.

4. For delamination, examination of such items as clad plate requires that a 90° angle probe be used.

5. For delamination in clad plate, a 100 percent scan of one surface is normally employed with a minimum 10 percent overlap. Rejection is generally based on a complete loss of back reflection that cannot be contained in a fixed diameter circle, usually 3 in. (7.5 cm). For

equipment being used in services that might generate hydrogen, the acceptance/rejection limits are often tightened so that defects that cannot be contained in a 1 in. (25 mm) circle are rejected.

6. To scan plate for laminations or other defects, a 90° probe is usually used for a 100 percent scan or a grid pattern may be established. With a grid, the plate is divided into a number of squares based on an arbitrary dimension, say 12-in. (30.5-cm) segments. Each square may be scanned, or a predetermined pattern of squares may be tested with the pattern being varied from plate to plate.
7. A couplant is required between the transducer and the work piece. The couplant has to have enough "body" to stay on the test surface during the test. Couplants that have to be replaced frequently during the test may prove unsatisfactory.
8. The material surface to be tested has to be smooth and flat enough to permit good probe contact. Tightly adherent mill scale and thin layers of smooth paint are satisfactory test surfaces. Weld spatter, dirt, loose scale, peeling paint, etc., produce unsatisfactory testing surfaces. The presence of sand, dirt, scale, etc., in the couplant material generally results in poor performance, so that couplant must be removed and replaced.
9. Calibration of an instrument is critical in flaw detection. To permit this calibration, various types of reflectors can be machined into blocks of metal similar to that being tested. Ideally, components such as tube, pipe, and plate, containing intentional defects of the type being sought, are used for calibration. Calibration method and the calibration standard are recorded. The operator's technique is recorded as well.
10. Photographs of screens showing defects may be used for a permanent record of test results. Alternatively, automatic scanning devices (C Scan) produce charts showing the results of the entire test. Many companies use C Scan only on critical or expensive equipment as it is relatively costly.

Ultrasonics is also used to determine the thickness of plate, heads, tube, pipe, etc. A short pulse of high frequency sound is sent into the pipe at intervals. When the sound strikes an interface such as the inside surface of the pipe, some of the sound echoes back to the transducer that receives the sound. The elapsed time between the pulse and echo is measured and converted into a height or distance on a screen, or as a digital readout (depending on the equipment used). Ultrasonic thickness is rapid and accurate but there are a number of difficulties that may arise.

The following are examples of some of the problems one may encounter.

1. The surface of the metal examined has to be smooth and clean. A rough or dirty surface can produce inaccurate readings, usually higher than the actual thickness.
2. The surface examined has to be relatively flat. A rocking probe can produce false thick or thin readings. Probe holders shaped to the component's curvature may be required to prevent rocking.
3. A clean, heavy-bodied couplant is required. Too light a couplant can also produce false readings.
4. The instrument has to be calibrated frequently as it may drift. Calibration on at least two thicknesses of material is generally recommended particularly if thickness variations in the material being tested are anticipated.
5. Also, low power from such things as run down batteries in the test instrument may produce false readings.

6. Readings are taken at several locations on a component, as the thickness may vary. An average of a minimum of three readings is considered to be the component thickness.

Magnetic Particle (MT) and Wet Flourescent Magnetic Particle Inspection (WFMP). Magnetic particle inspection, associated with the shop fabrication of pressure vessels and piping, is often employed on weld preparations to detect defects such as laminations, laps, folds, etc., that could open as welding progresses, resulting in unacceptable linear defects in the finished weld. Magnetic particle testing is also used to examine finished welds for surface and near-surface defects and to inspect raw materials such as bolts, nuts, valve components, flanges, and fittings for defects.

Magnetic particle inspection consists of inducing a magnetic field into a part, covering the part with ferromagnetic particles, and then examining it for signs of any discontinuity. If defects exist, localized magnetic fields are created. These localized fields attract magnetic particles which therefore outline the defect. Obviously, MT may be performed only on ferromagnetic material.

Magnetic particle testing may be divided into two general broad categories termed "dry" and "wet." The dry method signifies the application of dry iron powders to the surface under test; the wet method denotes the application of iron powders contained in a liquid. In addition, the particles in the wet suspension are fluorescent so that the tested surfaces can be examined under a black light. Dry powders may be of different colors to permit a contrast with the background, i.e., red particles would be used against a gray background. A popular variation involves painting the test area with a white lacquer to provide a contrast with the testing medium. As the iron particles in the liquid or suspension tend to be finer than those used with the dry method, the wet method is widely considered to be the more sensitive of the two.

Within these two classifications, wet and dry, MT may be identified by the type of equipment used: fixed (popularly known as a "wet bench"), prods, yoke, coils, etc. It is also designated as to the type of current, AC or DC, and to the type of magnetization, continuous or residual.

Fixed equipment such as the wet bench is used to examine relatively small components as they must be placed between fixed electrodes. A magnetic field is induced by applying a current while the product is flooded with a magnetic particle suspension. Examination of the product is then performed under a black light.

Prods are two hand-held electrodes that are pressed against a part to allow a magnetizing field to be generated in the part by the passage of the current. Prods are generally used on large or thick sections such as castings and are quite effective. However, poorly maintained or poorly applied prods cause arcing and will produce surface damage similar to welding electrode arc strikes. Many companies prohibit the use of prods on pressure equipment to avoid arcing damage.

Portable equipment such as yokes and coils are the most popular magnetic particle testing apparatus used in fabricating shops. Yokes may be AC or DC. AC yokes reveal only surface defects, but DC yokes will reveal subsurface as well as surface defects. Coils, which may be commercially produced products or simply cables wrapped around the component to be tested, will generally reveal surface and near-surface defects.

As with other testing methods, the skill of the operator influences not only the quality of the test but the interpretation of the results. Operators generally are qualified to ASNT, CGSB, or other suitable requirements. Again, many companies have in-house training programs to qualify their testing personnel without formally certifying them to recognized standards. Therefore, it may be necessary for the inspector to ask for a test on samples containing known defects using the proposed testing procedure before the fabrication starts.

The purchasing inspector once more requires a basic knowledge of the technology and some of the potential problem areas. There are a number of factors that may be monitored to ensure that the correct procedures and techniques are being followed. The following are some helpful suggestions.

1. Examine the test procedure to determine its adequacy. The procedure indicates, as a minimum, what equipment is to be used, the type of magnetizing current, the type of magnetization, (continuous or residual), the method (dry or wet), the direction of the field (must be a minimum of two mutually perpendicular directions), and any demagnetizing requirements.

2. Check the material to be tested to ensure that it is ferromagnetic. A simple magnet is adequate to determine ferromagnetism.

3. Check the surfaces to be tested to ensure that they are free of dirt, scale, grease, oil, etc., particularly for the wet method. Testing may be performed through paint with noncontact equipment. For contact equipment such as yokes, contact has to be made with bare metal to permit inducing a proper magnetic field.

4. For continuous magnetization, the component is examined while the magnetizing field is being applied.

5. With the dry method, ensure that excess powder is not used. Blowing gently on the part to remove excess powder is acceptable. The dry method is generally not used in the overhead position.

6. Make sure that MT is not being performed while welding, grinding, or some other operation is being performed on the component as this will affect the magnetic field.

7. The DC yoke is most effective for detecting subsurface defects. If repairs are performed, the DC yoke will reveal defects in areas that have been ground out. The AC yoke may not.

8. Permanent magnets may be used to induce a magnetic field, but many companies consider such a magnetic field insufficient to find anything but large defects.

9. Weld spatter, rough welds, porosity, undercut, excessive grinding, etc., will all effect the quality of the test. Linear indications may be revealed by rough grind marks, pitting, edges, rough weld beads, etc. Careful interpretation is required to separate defects from false indications.

10. The joint between a ferromagnetic material and a non-magnetic material will produce linear indications during magnetic particle testing.

11. After a linear indication is identified, viewing with a magnifying lens helps to identify the indication.

12. The closer the indication is to the surface, the sharper the particle pattern. Subsurface defects may be quite fuzzy, particularly with the wet method.

Liquid Penetrant Inspection (PT or LPT). Liquid penetrant inspection is probably the most widely used nondestructive technique in the fabrication of pressure equipment. It may be applied to both ferromagnetic and nonmagnetic materials to reveal surface defects. It does, however, require considerable surface preparation in that all grease, oil, dirt, scale, paint, etc., must be removed to permit penetration of the liquid or dye into potential defects.

The method involves the application of a penetrating dye that remains on the surface for a fixed period, known as the dwell time. This period is generally established according to the penetrant manufacturer's recommendations. Ten to fifteen minutes is a common dwell time.

After the dwell period, the excess penetrant is removed, generally by wiping with a clean cloth or similar material. For solvent removable penetrants, wiping with a cloth soaked in solvent is acceptable for removal. For water washable penetrants or post-emulsifying penetrants, the excess penetrant is removed with a low pressure water spray.

After the excess penetrant is removed, the surface being tested is developed using either a dry powder or a suspension of powder in water or solvent, depending on the system used. The surfaces are then examined.

Defects are revealed either by a colored dye bleeding out into the white developer for color contrast penetrants or under a black light as a fluorescent indication for fluorescent penetrants. With the color contrast penetrant, defects are relatively easy to see and photograph so a permanent record is possible.

Inspectors tend to be most familiar with this testing process because of its popularity and the relative ease in interpreting defects. Nevertheless, there are a several steps to ensure that proper testing is performed and they are as follows.

1. Examine the test procedure to confirm its adequacy. As a minimum, the procedure generally indicates:

    a) The brand name and specific type of penetrant, cleaner, emulsifier, and developer, as applicable.

    b) How the surfaces to be tested are to be cleaned and dried, including the drying time.

    c) How the penetrant is to be applied, how long it remains on the surface, and the temperature limitations, usually 60 to 125°F (15 to 52°C).

    d) How excess penetrant is to be removed and the surface dried before developing.

    e) How the developer is to be applied and the time before examination.

    f) How the tested surface is to be cleaned after the test.

2. Ensure that the surfaces to be tested are free of all oil, grease, scale, dirt, paint, etc., before testing.

    Welding smut on austenitic alloys welded with gas shielded processes is of particular concern as many fabricators claim this does not affect testing. It covers potential defects, absorbs the penetrant, and generally produces defect-like indications.

3. Ensure that the surfaces to be tested are dry before testing commences. Condensation may occur on cold components brought into a shop for testing. Therefore, warming to a 60°F (15°C) minimum may be necessary before testing.

4. Testing is performed away from grinding, sandblasting, painting, or other similar operations to prevent surface contamination and false indications.

5. Ensure that all penetrants have not exceeded their recommended shelf life before application.

6. All penetrants may require some form of agitation prior to application to prevent layering, separation, or other degradation.

7. Penetrant is applied by spray, brush, swab, or other suitable means and allowed to dwell for the specified period. Penetrant should remain wet for the dwell period so that additional applications during the dwell period may be required.

8. Excess penetrant is removed by wiping, water spray, solvent on cloth, etc., as applicable. Spraying solvent onto the test area to remove excess penetrant is not acceptable for the solvent system.

9. Ensure that the test areas are allowed to dry properly before developer is applied. Testing with a white cloth may be necessary.

10. Ensure that the developer is properly mixed to provide adequate dispersion of the suspended particles.

11. Water suspension developers are applied by dipping or spraying to obtain a thin coating over the entire surface being tested.

12. Solvent suspension developers are applied by spraying.

13. Ensure that the developer is not too thick or too thin. Insufficient developer will not draw out the penetrant and too much developer may mask defects.

14. The tested surface is generally examined as the developer is applied to note any gross defects from which the penetrant bleeds profusely. Final interpretation is normally performed after the penetrant has been allowed to bleed for 7 to 30 minutes. For large surfaces, development and examination by sections may be required to prevent exceeding recommended bleed times.

15. Localized surface irregularities such as machine marks, rough welds, undercut, etc., may produce false indications, so careful interpretation is required. Scale filled defects may not be revealed.

16. Broad areas of fluorescence or pigmentation, which may mask defects, are unacceptable. Such areas are cleaned and retested.

References

1. American Society of Mechanical Engineers, ASME Boiler and Pressure Vessel Codes, Section VIII, Appendix 10, "Quality Control System" (New York, NY: ASME).

2. ASTM A 262, "Standard Practices for Detecting Susceptibility to Intergranular Attack in Austenitic Stainless Steels" (Philadelphia, PA: ANSI).

3. American National Standards Institute, ANSI/ASME B1.20.1, "Pipe Threads, General Purpose"(New York, NY: ANSI).

4. American National Standards Institute, ANSI/ASME B31.1, "Power Piping" (New York, NY: ANSI).

5. American National Standards Institute, ANSI/ASME B31.3, "Chemical Plant and Petroleum Refinery Piping" (New York, NY: ANSI).

6. American National Standards Institute, ANSI/ASME B31.2, "Fuel Gas Piping" (New York, NY: ASME).

7. American Petroleum Institute, API 5CT, "Specification for Casing and Tubing (US Customary Units)" (Washington, DC: API).

   American Petroleum Institute, API 5CTM, "Specification for Casing and Tubing (Metric Units)" (Washington, DC: API).

8. American Petroleum Institute, API 5L, "Specification for Line Pipe" (Washington, DC: API).

9. American National Standards Institute, ANSI/ASME B36.10, "Welded and Seamless Wrought Steel Pipe" (New York, NY: ANSI).

10. American National Standards Institute, ANSI/ASME B36.19, "Stainless Steel Pipe" (New York, NY: ANSI).

11. ASTM A 53, "Standard Specification for Pipe, Steel, Black and Hot-Dipped, Zinc-Coated, Welded and Seamless" (Philadelphia, PA: ASTM).

12. ASTM A 106, "Standard Specification for Seamless Carbon Steel Pipe for High Temperature Service" (Philadelphia, PA: ASTM).

13. Pipe Fabrication Institute, PFI ES-24-84, "Pipe Bending Methods, Tolerances, Process, and Material Requirements" (Springdale, PA: PFI).

14. ASTM A 587, "Standard Specification for Electric-Welded Low-Carbon Steel Pipe for the Chemical Industry" (Philadelphia, PA: ASTM).

15. NACE International Standard MR0175, "Sulfide Stress Cracking Resistant Metallic Materials for Oilfield Equipment" (Houston, TX: NACE).

16. American Society of Mechanical Engineers, ASME Boiler and Pressure Vessel Codes, Section IX, "Qualification Standard for Welding and Brazing Procedures, Welders, Brazers, and Welding and Brazing Operators" (New York, NY: ASME).

17. American National Standards Institute, ANSI/AWS A2.4, "Standard Symbols for Welding, Brazing, and Nondestructive Examination" (New York, NY: ANSI).

18. American Society of Mechanical Engineers, ASME Boiler and Pressure Vessel Codes, Section V, "Nondestructive Examination" (New York, NY: ASME).

19. American Society of Mechanical Engineers, ASME Boiler and Pressure Vessel Codes, Section VIII D1, "Rules for Construction of Pressure Vessels" (New York, NY: ASME).

Definitions

The following definitions were extracted from the CSA[10] Z299 Program and from ANSI/ASQC[11] A3, "Quality Systems Terminology."

Audit A documented activity aimed at verifying, by independent examination and evaluation, that the applicable elements of the quality program have been established, documented, and implemented in accordance with contract requirements.

Batch or Lot An identifiable collection of products or quantity of items of a single type, grade, class, size, or composition produced in the same plant, under essentially the same conditions, and at essentially the same time.

Calibration The comparison of two instruments or measuring devices, one of which is of known accuracy traceable to a recognized national or international standard, to eliminate any variation in accuracy among instruments or measuring devices.

Characteristic Any distinct property or attribute of a product, process, or service that can be described and measured to determine compliance with specified requirements.

Contract The written covenant and other documents agreed to and legally binding between the customer and supplier that outline requirements and conditions that must be met to successfully complete the work.

Customer The party or representative issuing a contract for procuring goods or services.

Design Input Requirements and information outlined by the customer, regulatory authorities, design engineers, or other engineering disciplines that form the basis for the design of a component.

Design Output Requirements and information needed to procure or fabricate products.

Design Review A formal, independent examination of a design to confirm its adequacy.

Disposition An action to determine how a nonconformance is to be resolved.

---

[10]Canadian Standards Association, Rexdale, Ontario, Canada.
[11]American Society for Quality Control, Milwaukee, WI.

Documentation Written records, procedures, test reports, drawings, design calculations, and other evidence that determine if a component meets with the requirements of the contract.

Evaluation An appraisal used to determine whether or not a product, company, or quality program is capable of meeting customer requirements.

Inspection The examination, measurement, and testing of the characteristics of products or services to determine their conformance with contract requirements.

Inspector The person employed or contracted by the fabricator or supplier who measures, examines, and tests components to ensure their compliance with the contract.

Inspection and Test Plan A document describing how, when, where, and how often a company will inspect or test a product to provide objective evidence that the product meets the contract specification.

Inspection Hold Point A point in the manufacturing process beyond which fabrication may not proceed until the required inspection has been performed and documented.

Inspection Witness Point A point in the manufacturing process beyond which fabrication may not proceed until the required inspection has been performed by the fabricator and witnessed by the customer, regulatory agency, or other authority identified in the Inspection and Test Plan.

Nonconformance A deficiency in a characteristic, documentation, fabrication, or test procedure that results in the quality of the product or service being unacceptable or indeterminant in terms of the specified requirements.

Objective Evidence Any recorded result of measurements, tests, observations, or other means of positively verifying that product quality meets contract requirements.

Poison Pad An alloy pad welded to a high-alloy vessel shell. Low-alloy attachments are welded to the pad, preventing contamination of the shell with low-alloy material.

Positive Recall A method whereby a component may be released for further work even though its quality has not been verified. The product is identified to permit its removal for correction if it proves unacceptable.

Quality Assurance All planned and systematic activities needed to provide adequate confidence that the product or service meets contract requirements.

Quality Control The means employed by the fabricator to ensure compliance with the contract requirements, including production of a quality manual, inspection activities, and the production of required documentation.

Quality Management	The overall management function that determines and implements quality policy.
Quality Policy	The overall intentions and directions of an organization regarding quality, as expressed by senior management.
Quality Program	All the methods, tests, procedures, specifications, and documentation used by the fabricator to ensure compliance with the contract quality requirements.
Quality System	The organizational structure, responsibilities, procedures, processes, and resources employed to implement quality management.
Regulatory Authority	Federal, state, provincial, territorial, or municipal agency having the lawful right to impose laws and regulations on the fabrication, testing, and operation of pressure components.
Repair	The correction of defects so that the components can function reliably and safely in service.
Rework	Reprocessing components so that they conform to the original specifications
Special Inspection Processes	An inspection requiring either specialized inspector skills, specialized techniques, or both.
Subcontract	A contract between the prime contractor or supplier and a sub-supplier.
Supplier	The party responsible for the performance of work outlined in the contract.
Surveillance	The continuing evaluation and analysis of records, methods, procedures, material, and work functions to assure conformance with contract requirements.
Surveyor	The customer's representative who carries out the surveillance function.
Verification	Examination and testing carried out by the surveyor to verify that materials conform to the contract requirements.
Work	All labor, materials, equipment, services, supplies, and actions required to be furnished or performed by the contractor to meet contract requirements.

Nuts and Bolts

The chemical division of American Cyanamid Company, now Cytec Industries Inc.,[12] recognizing that the identification and proper application of fasteners affects the integrity of pressure systems, has produced a video on the subject titled "Nuts and Bolts." Permission was granted by American Cyanamid to reproduce the companion book to the video series in an effort to assist inspectors in both identifying fasteners, and in determining the correctness of the intended service.

TYPES OF BOLTS AND SCREWS

BOLTS

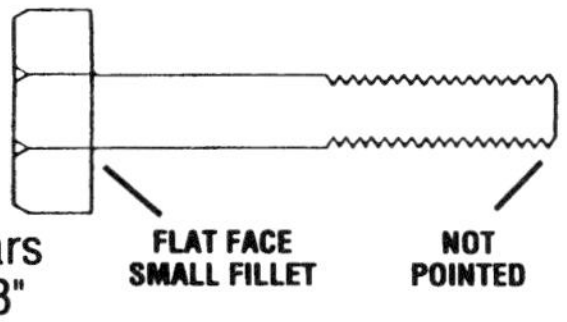

- HEX BOLT - as shown in drawing.
- HEAVY HEX BOLT - appears the same except that it is 1/8" longer across the flat of the head than a hex bolt.
- SQUARE BOLT - appears the same except that the head is square.

Except for 1/4" bolts, both the hex and square bolts take the same wrench size.

"Use bolts only with nuts. Screws with nuts or drilled & tapped holes."

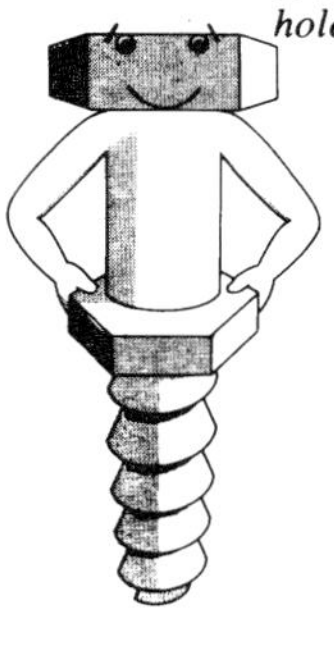

SCREWS

- HEX CAP SCREW (finished hex bolt) - as shown in drawing.
- HEAVY HEX SCREW - appears the same except it is 1/8" longer across the flat of the head than a hex cap screw.
- HEAVY HEX STRUCTURAL BOLT - appears the same as a heavy hex screw except that the shank is much longer for the same length fastener.

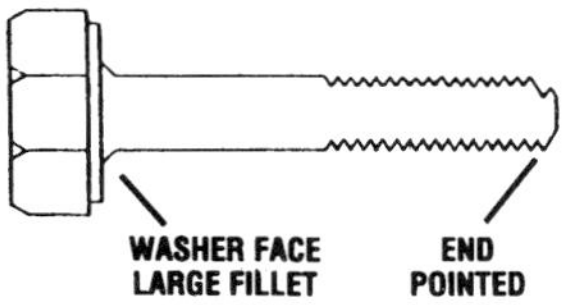

LOW STRENGTH FASTENERS

For use with soft gaskets and/or brittle flanges and in flanges of Class 300 and lower pressure rating. (Soft gaskets include most rubber and sheet plastic gaskets.) (Brittle flanges include cast iron, lined steel and plastic flanges.) NOTE: The symbol @ indicates letters or symbols used for manufacturer's identification. Such markings may not be on ASTM A307 fasteners. All studs made to ASTM A307, A193 and A320 have the same markings on one end as the heads of bolts or screws.

ASTM A307 Grade B

- Plain or galvanized standard carbon steel flange fastener for use from -20 to 450°F. TYPE - Heavy Hex Bolt. NUTS - either ASTM A563 Grade A or ASTM A194 Grade 2H.

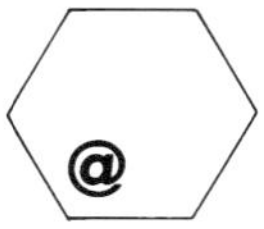

ASTM A307 Grade A or SAE Grade 2

- Plain or galvanized standard carbon steel machinery fastener for use from -20 to 450°F. DO NOT use with brittle flanges. TYPE - Hex Bolt. NUTS - either ASTM A563, Grade A, ASTM A194 Grade 2H or SAE Grade 2.

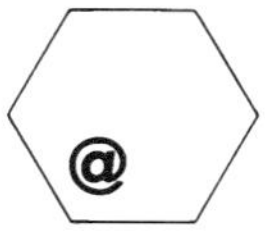

---

[12] Cytec Industries Inc.
10800 River Road
Westwego, LA 70094.

LOW STRENGTH FASTENERS

ASTM A449 or SAE Grade 5

• Medium carbon or low alloy plain or galvanized steel fasteners used for machinery service from -20 to 450°F; not normally used for flange service. SAE Grade 5 fasteners are common in both fine and coarse thread. TYPE - Hex Bolt and Hex Cap Screw. NUTS - Either ASTM A563 Grade C, ASTM A194 Grade 2H or SAE Grade 5.

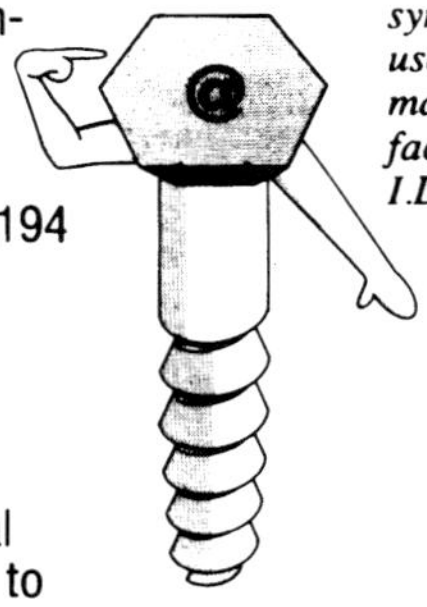

"The @ indicates letters or symbols used for manufacturer's I.D."

ASTM A325 Type 1

• Fastener used in structural steel joints. Material identical to ASTM A449 and nuts are also the same. TYPE - Heavy Hex Structural Bolt either plain or galvanized with coarse threads. There are two other types of A325 fasteners but they are uncommon.

INTERMEDIATE STRENGTH FASTENERS

For use with hard gaskets and service with Class 300 flanges and higher ratings. (Hard gaskets include hard sheet gaskets and spiral wound gaskets).

ASTM A193 Grade B7M

• Alloy steel fastener with controlled maximum hardness for H_2S, HF and HCN service between -20 to 1000°F. TYPE - Heavy Hex Screw. Use only A194 Grade 2HM nuts. For service from -100 to -20°F fasteners per ASTM A320 Grade L7M should be used. These fasteners will be marked "L7M". NUTS - ASTM A194 Grade 7M.

"Hard gaskets include hard sheet gaskets and spiral wound gaskets."

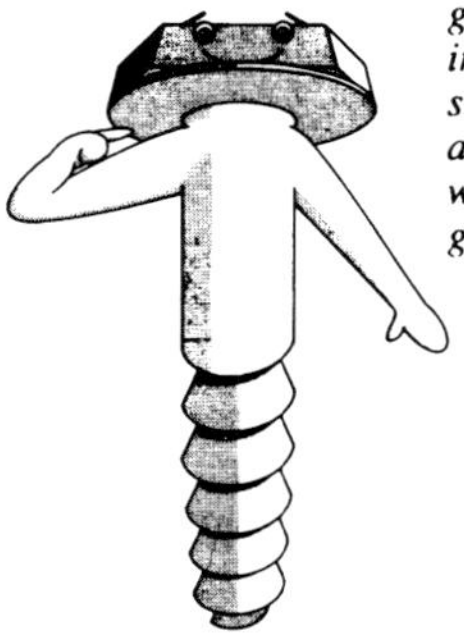

LOW STRENGTH FASTENERS

ASTM A193 Grade B8

• Standard Type 304 stainless steel flange fastener for use from -20 to 1500°F. TYPE - Heavy Hex Screw. NUTS - ASTM A194 Grade 8.

Type 304 S/S, Grade B8
Nuts - Grade 8

• Type 316 S/S is marked B8M: NUTS - A194 Grade 8M.

• Type 321 S/S is marked B8T: NUTS - A194 Grade 8T.

There are several other less common Type 300 series stainless grades. For service from -325 to -20°F fasteners per ASTM A320 should be used; stainless steel ASTM A320 fasteners are marked the same as ASTM A193.

Type 316 S/S, Grade B8M
Nuts - Grade 8M

Type 321 S/S, Grade B8T
Nuts - Grade 8T

TYPE 304 Stainless Steel

• These fasteners may or may not by made to ASTM F593 Group 1. The alloy may not meet the chemical requirements for Type 304 S/S. TYPE - Hex Cap Screw. SHOULD NOT BE USED AS FLANGE FASTENERS.

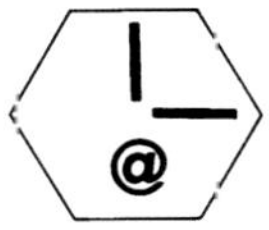

INTERMEDIATE STRENGTH FASTENERS

ASTM A193 Grade 8, Class 2

• Strain harden Type 304 stainless steel flange fasteners for service from -20 to 800°F.
TYPE - Heavy Hex Screw.
NUTS - ASTM A194 Grade 8.

Strain harden
Type 304 S/S

TYPE 316 S/S is marked "B8M2":
NUTS - A194 Grade 8M.

Strain harden
Type 316 S/S

TYPE 321 S/S is marked "B8T":
NUTS - A194 Grade 8T.

Strain harden
Type 321 S/S

• There are several other less common Type 300 series stainless steel grades. For service from -325 to -20°F fasteners per ASTM A320 should be used. ASTM A320 stainless steel fasteners are marked the same as in ASTM A193.

HIGH STRENGTH FASTENERS

Use with hard gaskets to include metal gaskets and Class 300 and higher pressure flanges. DO NOT USE IN H_2S,HF OR HCN SERVICE.

ASTM A193 Grade B7

• Alloy steel fastener for service from -20 to 1000°F. Standard fastener for steam service. TYPE - Heavy Hex Screw. NUTS - ASTM A194 Grade 2H.

ASTM A320 Grade L7

• Low temperature fastener similar to ASTM A193 Grade B7 for use at -150 to 100°F. TYPE - Heavy Hex Screw. NUTS - ASTM A194 Grades 4 or 7.

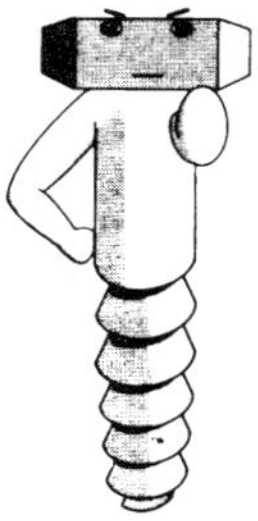

"Do not use in H_2S, HF or HCN service."

VERY HIGH STRENGTH FASTENERS

Rarely recommended for use in flanges but often necessary in machinery. Corrosion may result in embrittlement. DO NOT galvanize or couple to galvanized surfaces or use in corrosive service especially H_2S,HF and HCN.

ASTM A354 Grade BD and SAE Grade 8

• Plain alloy steel fasteners for service from -20 to 450°F in machinery. DO NOT use in flanges. TYPE - Hex Cap Screw or Hex Bolt with either fine or coarse threads. NUTS - Either ASTM A563 Grade DH or SAE Grade 8.

SAE Grade 8.2

• Special low carbon steel with room temperature properties similar to SAE Grade 8. Usage, type and suitable nuts are the same as SAE Grade 8.

METRIC FASTENERS

Metric fasteners made to the requirements of ASTM A193M, A194M and A320M are marked the same as the corresponding fasteners made to ASTM A193, A194 and A320. Metric fastener grades that correspond to the standard SAE grades are made to the requirements of ASTM F568 and may be marked as shown below, but the marking of grade and manufacturer's identification is optional.

SAE Grade 2
Metric Grade 4.6

SAE Grade 5
Metric Grade 8.8

SAE Grade 8
Metric Grade 10.9

OTHER GRADES OF FASTENERS ARE AVAILABLE BUT THEIR USE IN CHEMICAL PLANTS IS RARE. SOME OF THESE FASTENERS MAY NOT BE SUITABLE FOR THE INTENDED SERVICE; CHECK WITH YOUR FOREMAN AS TO THEIR SUITABILITY.

POINTS TO REMEMBER

- ***Use the correct nut and bolt for the application.***
- ***Only use clean, well-lubricated fasteners.***
- ***Use the correct wrench.***
- ***Tighten the fastener to the correct torque.***
- ***Follow the manufacturer's recommended torque requirement.***
- ***Replace fasteners with the same grade.***
- ***Know how to identify the various grades of fasteners.***
- ***Seek engineering assistance to determine correct grades of fasteners and to determine causes of fastener failures.***
- ***Use cap or machine screws or pointed studs in a drilled and tapped hole.***

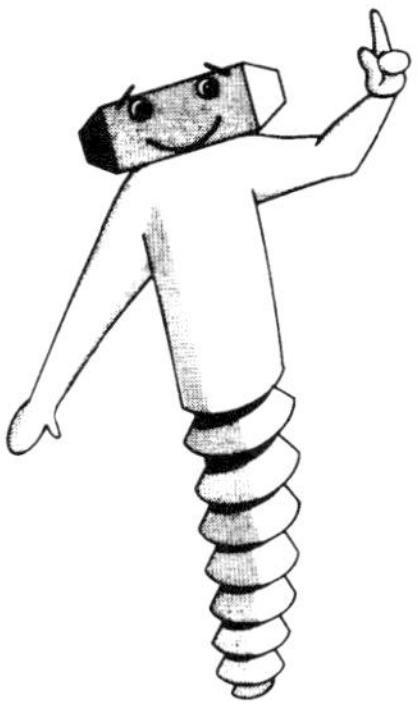

THINGS TO AVOID

• Do not use incorrect grades of fasteners.
• Do not reuse damaged fasteners.
• Do not use too large or too small a wrench for the application.
• Do not assume that all fasteners are the same.
• Do not over torque or under torque fasteners.
• Do not weld steel fasteners between the head and nut or between both nuts in a stud.
• Do not use bolts in a drilled and tapped hole.
• Do not ignore repeated fastener failures.
• Do not use stainless steel fasteners with steel or cast iron flanges, especially for service above 200°F.

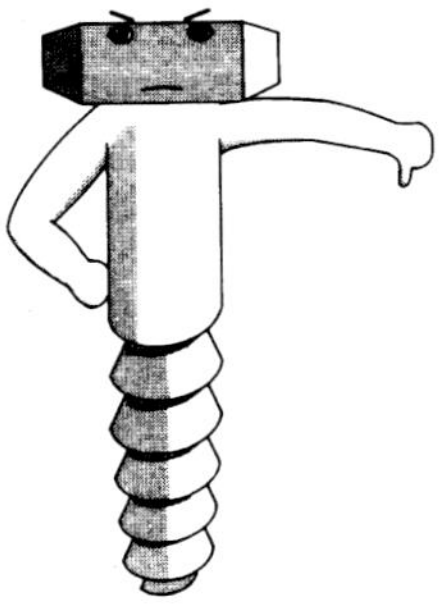

FASTENER TORQUE

• Torque tables are only accurate if the fasteners are clean and well lubricated.
• Torque tables are usually written for machinery applications.
• If the manufacturer recommends a different torque than other sources, use the manufacturer's recommended torque.
• Too little torque may result in fatigue failure.

TORQUE TABLE FOR MACHINERY, HEX CAP SCREWS, AND COARSE THREAD

FASTENER SIZE	SAE Grade 2	SAE Grade 5	SAE Grade 8
1/4"	3 ft. lbs.	5 ft. lbs.	7 ft. lbs.
5/16"	7	10	15
3/8"	12	19	26
7/16"	19	30	42
1/2"	29	45	64
9/16"	43	66	92
5/8"	58	90	125
3/4"	100	160	225
7/8"	100*	260	365
1"	150*	390	545

** Strength level of steel is lower for these fastener sizes.*

• Gasketed Flange Surfaces - Torque value is between 15% and 80% of the value given in the Table. Use low torque values for soft gaskets or brittle flanges. Use high torque value for hard gaskets. Plastic flange manufacturers publish recommended torque values.
• Structural Bolting - Use 1.33 times the torque value given for SAE Grade 5 fasteners.

Heat Exchanger Check Sheets

The following check, or data, sheets of a major chemcial company outline a comprehensive inspection program for metallic heat exchangers. They may be used as a guide or model ofr inspectors who must prepare such documents. The data sheets when completed form a summary record of fabrication inspection. Of course, all items on the sheets may not be required for every unit. The inspector will complete only the applicable items using the notations N.A. or N/A for non-applicable items.

INSPECTION DATA SHEET (IDS)
HEAT EXCHANGERS METALLIC

DATE (Y-M-D) | REV.

PAGE 1 OF 9

1	PO. No.	EQUIPMENT No.				
2	DESCRIPTION	LOCATION				
3	PROJECT	INSPECTION BY Initial and date when complete. X = witness ● = hold point o = spot				
4	PROJECT No.					
5	VENDOR					
6	VENDOR LOCATION					
7	CODES/STANDARDS	VENDOR		PURCHASER		
8						
9	PURCHASERS SPEC No. REV.	Initial	Date	Initial	Date	
10	DRAWING No. REV.					
11						
12						
13	FOR NOTES AND COMMENTS SEE PAGE					
14						
15	THE QA/QC PROGRAM SHALL INCLUDE BUT IS NOT LIMITED TO					
16	THE FOLLOWING :					
17						
18						
19	INSPECTION - FABRICATION SCHEDULE					
20	Revised Weekly ☐ Monthly ☐					
21						
22	INSPECTION AND TEST PLAN					
23						
24	DOCUMENTATION APPROVAL BY REGULATORY BODY PRIOR TO					
25	FABRICATION :					
26	QA Manual					
27	Fabrication Drawings					
28	Design Calculations					
29	Welding Procedures (WPS)					
30	Welding Procedure Qualification Records (PQR)					
31	Material Test Certificates					
32	Nondestructive Testing Procedures					
33	Heat Treatment Procedures					
34	Pressure Testing Procedures					
35						
36						
37	DOCUMENTATION REVIEW BY PURCHASER PRIOR TO					
38	FABRICATION :					
39	QA Manual					
40	Fabrication Drawings					
41	Design Calculations					
42	Welding Procedures (WPS)					
43	Welding Procedure Qualification Records (PQR)					
44	Welder's Qualification Records					
45	Material Test Certificates					
46	Nondestructive Testing Procedures					
47	Heat Treatment Procedures					
48	Pressure Testing Procedures					
49						
50						
51						
52						

PURCHASER	VENDOR
QA APPROVED	APPROVED
Print Signature Date (Y-M-D)	Print Signature Date (Y-M-D)

INSPECTION DATA SHEET (IDS)
HEAT EXCHANGERS METALLIC

DATE (Y-M-D) | REV.

PAGE 2 OF 9

INSPECTION BY
Initial and date when complete.

X = witness ● = hold point o = spot

					Vendor Initial	Vendor Date	Purchaser Initial	Purchaser Date
53	**EQUIPMENT No.**							
54	**DESCRIPTION**							
55								
56								
57								
58								
59	**MATERIALS**							
60	Material Test Certificates							
61	Mechanical Properties							
62								
63								
64	**TESTS**	**IMPACT**	**UT**	**PMI**				
65	Shell	☐	☐	☐				
66	Tubesheet/s	☐	☐	☐				
67	Channel – Inlet	☐	☐	☐				
68	Channel – Outlet	☐	☐	☐				
69	Shell Cover/s	☐	☐	☐				
70	Floating Head Cover	☐	☐	☐				
71	Channel Nozzles – Inlet	☐	☐	☐				
72	Channel Nozzles – Outlet	☐	☐	☐				
73	Shell Nozzles	☐	☐	☐				
74	Shell Flanges	☐	☐	☐				
75	Channel Flanges – Inlet	☐	☐	☐				
76	Channel Flanges – Outlet	☐	☐	☐				
77	Bolts or Studs	☐	☐	☐				
78	Nuts	☐	☐	☐				
79								
80								
81	Clad Plates	☐	☐	☐				
82	Clad Plates – Shear Test							
83								
84								
85	Tubes							
86	Ultrasonic Test							
87	Eddy Current Test							
88	Air Under Water Test							
89	Tightness Test of U-Bends							
90	Heat Treatment of U-Bends							
91	Corrosion Testing							
92								
93								
94	Materials Traceability or Marking:							
95	Shell							
96	Tubesheet/s							
97	Channel/s							
98	Cover/s							
99	Head/s							
100	Tubes							
101	Baffles							
102	Tie Rods							
103	Partition Plates							
104	Nozzles							

PURCHASER
QA APPROVED

Print Signature Date (Y-M-D)

VENDOR
APPROVED

Print Signature Date (Y-M-D)

INSPECTION DATA SHEET (IDS) HEAT EXCHANGERS METALLIC

DATE (Y-M-D) | REV.

PAGE 3 OF 9

		INSPECTION BY			
105	**EQUIPMENT No.**	Initial and date when complete.			
106	**DESCRIPTION**				
107		X = witness • = hold point o = spot			
108					
109		VENDOR		PURCHASER	
110					
111	Materials Traceability or Marking (Continued)	Initial	Date	Initial	Date
112	Flanges				
113	Bolts or Studs				
114	Nuts				
115	Lifting Lugs				
116	Supports				
117	Gaskets				
118					
119	Welding Consumables				
120	Welding Fluxes ☐ Purge Gases ☐				
121					
122					
123	**WELDING**				
124	Welding Procedure Specification (WPS)				
125	Procedure Qualification Record (PQR)				
126	Welder's Qualification Records				
127	Corrosion Testing				
128	Weld Preparations				
129	Dimensional Exam.				
130	Visual Exam.				
131	Magnetic Particle Exam. (MT)				
132	Liquid Penetrant Exam. (PT)				
133	Ultrasonic Exam. (including nozzle openings) (UT)				
134	Nozzles and Pads – Fit up				
135	Back Gouging				
136	Grinding Welds Flush				
137	Channels/s Inside ☐ Outside ☐				
138	Shell Inside ☐ Outside ☐				
139					
140					
141	**TUBE-TO-TUBESHEET WELDING**				
142	Welding Procedure				
143	Welding Procedure Qualification				
144	Simulated Post Weld Heat Treatment				
145	Sectioning and Examination of Test Welds				
146	Corrosion Testing ☐ Hardness Testing ☐				
147	Welder's Qualification Records				
148	Procedure Qualification Record				
149	Tube Pattern ☐ Number of Holes ☐				
150	Tube Hole Size ☐ Pitch ☐				
151	Tube Hole Finish				
152	Tube Hole Grooves and Size				
153	Tube Protrusion				
154	Tube flush with Tubesheet				
155					
156	Tube Expansion – Before ☐ After ☐ Welding				

PURCHASER	VENDOR
QA APPROVED	**APPROVED**
Print _____ Signature _____ Date (Y-M-D) _____	Print _____ Signature _____ Date (Y-M-D) _____

INSPECTION DATA SHEET (IDS)
HEAT EXCHANGERS METALLIC

DATE (Y-M-D) | REV.

PAGE 4 OF 9

INSPECTION BY

Initial and date when complete

X = witness • = hold point o = spot

No.	Item	VENDOR Initial	VENDOR Date	PURCHASER Initial	PURCHASER Date
157	**EQUIPMENT No.**				
158	**DESCRIPTION**				
159					
160					
161					
162					
163	**TUBE-TO-TUBESHEET WELDING** (Continued)				
164	Tube Pull-out Tests				
165					
166	Tube Diameter				
167	Tube Wall Thickness				
168					
169	Cleanliness Tubes ☐ Holes ☐				
170	Tube Pickling ☐ Passivating ☐				
171	Dye Penetrant Test ☐ After first weld pass				
172	☐ After every weld pass				
173	☐ After final weld pass				
174					
175	Post Weld Heat Treatment (PWHT)				
176	Radiography Before PWHT ☐ After PWHT ☐				
177	Magnetic Particle Exam. Before PWHT ☐ After PWHT ☐				
178					
179					
180	**TUBE EXPANDING**				
181	Procedure Specification				
182	Procedure Qualification				
183	Location and Size of Grooves				
184	Tube Hole Finish				
185	Tube Protrusion				
186	Length of Expansion				
187	Expansion Dry ☐ Lubricated ☐				
188	Expansion before welding				
189	Expansion after welding				
190					
191					
192	**VISUAL EXAMINATION**				
193	Tubes Inside ☐ Outside ☐				
194	Plates and Other Components				
195	Channels Inside ☐ Outside ☐				
196	Shell Inside ☐ Outside ☐				
197					
198					
199	NOTE: If shell or channel is lined or coated				
200	complete separate IDS				
201					
202					
203	**NONDESTRUCTIVE TESTING OF SHELL, CHANNEL, NOZZLE**				
204	**AND ATTACHMENT WELDS -WHEN ACCESSIBLE**				
205	Nondestructive Testing Procedures				
206	Ultrasonic Exam.				
207	After welding ☐ After PWHT ☐				
208					

PURCHASER	**VENDOR**
QA APPROVED	**APPROVED**
Print Signature Date (Y-M-D)	Print Signature Date (Y-M-D)

INSPECTION DATA SHEET (IDS)
HEAT EXCHANGERS METALLIC

DATE (Y-M-D) | REV.

PAGE 5 OF 9

209	**EQUIPMENT No.**	**INSPECTION BY**			
210	**DESCRIPTION**	Initial and date when complete			
211					
212		X = witness ● = hold point o = spot			
213					
214		**VENDOR**		**PURCHASER**	
215	**NON-DESTRUCTIVE TESTING OF SHELL, CHANNEL, NOZZLE**	Initial	Date	Initial	Date
216	**AND ATTACHMENT WELDS -WHEN ACCESSIBLE** (Continued)				
217	Radiographic Exam.				
218	After welding ☐ After PWHT ☐				
219	Magnetic Particle Exam.				
220	After welding Inside ☐ Outside ☐				
221	After PWHT Inside ☐ Outside ☐				
222	Dye Penetrant Exam				
223	After welding Inside ☐ Outside ☐				
224	After PWHT Inside ☐ Outside ☐				
225					
226	Spectrographic Testing before PWHT				
227	Base Metal				
228	Weld Metal Inside ☐ Outside ☐				
229					
230					
231	**HEAT TREATMENT**				
232	Procedures				
233	Equipment Calibration				
234	Heat Treatment of:				
235	Shell				
236	Channel/s				
237	Cover/s				
238	Head/s				
239	Tubesheet/s				
240	Tube to Tubesheet Welds One side ☐ Both sides ☐				
241	Heat Treatment Charts – Records				
242	Confirm no welding after stress relieving				
243					
244					
245	**HARDNESS TESTING OF WELDS**				
246	Procedure				
247	Equipment Calibration				
248	Tube Side after welding Inside ☐ Outside ☐				
249	Tube Side after PWHT Inside ☐ Outside ☐				
250					
251	Shell Side after welding Inside ☐ Outside ☐				
252	Shell Side after PWHT Inside ☐ Outside ☐				
253					
254					
255	**DIMENSIONAL TESTING**				
256	Tubes:				
257	Length				
258	Diameter				
259	Wall Thickness				
260					

PURCHASER	VENDOR
QA APPROVED	**APPROVED**
Print ____ Signature ____ Date (Y-M-D) ____	Print ____ Signature ____ Date (Y-M-D) ____

INSPECTION DATA SHEET (IDS)
HEAT EXCHANGERS METALLIC

DATE (Y-M-D) | REV.

PAGE **6** OF **9**

No.		INSPECTION BY Initial and date when complete. X = witness ● = hold point o = spot			
		VENDOR		PURCHASER	
261	**EQUIPMENT No.**				
262	**DESCRIPTION**				
263					
264					
265					
266					
267	**DIMENSIONAL TESTING** (Continued)	Initial	Date	Initial	Date
268	Tubesheets				
269	Plate Thickness				
270	Cladding Thickness				
271	Diameter				
272	Bolt Hole Size ☐ Circle Layout ☐				
273	Tube Hole Pattern Tubesheet ☐ Baffles ☐				
274	Tube Hole Diameter Tubesheet ☐ Baffles ☐				
275	Tube Pitch ☐ Ligaments ☐				
276					
277	Tube Bundle				
278	Baffles Type ☐ Number ☐				
279	Tierods ☐ Spacers ☐				
280					
281	Shell and Channels				
282	Plate Thickness				
283	Plate Alignment				
284	Inside Diameter				
285	Roundness				
286	Length				
287	Shell Cover				
288	Head/s				
289	Channel Cover/s				
290					
291	Nozzles and Openings				
292	Orientation of Nozzles				
293	Length ☐ Diameter ☐				
294	Flange Type ☐ Rating ☐				
295	Flange Gasket Face Finish				
296	Partition Plate				
297	Impingement Plate/s				
298					
299	Radius Tubesheet to Channel ☐ Shell ☐				
300	Bolting Flange ☐ Internal ☐				
301					
302					
303	**GENERAL TESTS**				
304	Weld Production Test Plates				
305					
306	Tube-to-Tubesheet Welds:				
307	Tightness test with media as specified				
308	After 1st pass ☐ 2nd pass ☐ Final pass ☐				
309					
310	Hydrostatic Pressure Test Shell Side				
311	Tube Side				
312					

PURCHASER	VENDOR
QA APPROVED	**APPROVED**
Print ____ Signature ____ Date (Y-M-D) ____	Print ____ Signature ____ Date (Y-M-D) ____

INSPECTION DATA SHEET (IDS)
HEAT EXCHANGERS METALLIC

DATE (Y-M-D) | REV.

PAGE 7 OF 9

EQUIPMENT No.

DESCRIPTION

INSPECTION BY
Initial and date when complete.
X = witness • = hold point o = spot

	VENDOR		PURCHASER	
	Initial	Date	Initial	Date
GENERAL TESTS (Continued)				
Water Quality ☐ Temperature ☐				
Differential Pressure Test				
Leakage Test:				
Air ☐ Freon ☐ Helium ☐				
Shell Side				
Tube Side				
Recording of Pressure Test				
FINAL INSPECTION				
Conformity with Order Documents				
Material's Stamps and Traceability				
Welds and Weld Spatter Cleaning				
Cleanliness Inside ☐ Outside ☐				
Pickling and Passivating				
Vacuum Drying				
Painting or Coating				
Paint or Coat Thickness				
Corrosion Protection				
Dimensional Examination				
Completeness				
Name Plate				
SPARE PARTS				
Bolts ☐ Nuts ☐				
Gaskets				
DOCUMENTATION REQUIRED FOR RELEASE FOR SHIPMENT				
Vendor's Completed IDS				
Final Fabrication Drawings				
Welding Procedure Specifications (WPS)				
Welding Procedure Qualification Records (PQR)				
Welder Performance Qualification Records				
Weld Map				
NDT Procedures				
NDT Personal Qualifications				
Pressure Test Chart ☐ Certificate ☐				
Leak Test Certificate				

PURCHASER	VENDOR
QA APPROVED	**APPROVED**
Print ____ Signature ____ Date (Y-M-D)	Print ____ Signature ____ Date (Y-M-D)

INSPECTION DATA SHEET (IDS)
HEAT EXCHANGERS METALLIC

DATE (Y-M-D) | REV.

PAGE 8 OF 9

		INSPECTION BY			
365	**EQUIPMENT No.**	Initial and date when complete			
366	**DESCRIPTION**				
367					
368		X = witness ● = hold point o = spot			
369		VENDOR		PURCHASER	
370					
371	**DOCUMENTATION REQUIRED FOR RELEASE FOR**	Initial	Date	Initial	Date
372	**SHIPMENT** (Continued)				
373	Heat Treatment Chart ☐ Certificate ☐				
374					
375	Material Test Certificates				
376	Material Traceability				
377	Positive Material Identification (PMI) Report				
378	Ultrasonic Exam. Plan and Reports				
379	Radiographic Exam. Plan and Reports				
380	Magnetic Particle Exam. Plan and Reports				
381	Dye Penetrant Exam. Plan and Reports				
382	Hardness Test Reports				
383					
384	Copy of Nameplate Rubbing				
385	Serial Number				
386	Affidavit of Manufacture				
387					
388					
389	**SHIPPING**				
390	Internal Dryness				
391	Protective Coating of Flange Faces				
392	Protection of Machined Surfaces				
393	Vent Holes plugged with Grease				
394	Bracing and Support of Load				
395	Packing and Marking				
396	Packing List				
397	Nitrogen Atmosphere for Shipping				
398					
399	Routing and Shipping Clearance				
400	Custom Clearance				
401					
402					
403	**RELEASE FOR SHIPMENT**				
404	Upon receipt of copy of completed IDS				
405					
406					
407					
408					
409					
410					
411					
412					
413					
414					
415					
416					

PURCHASER			VENDOR		
QA APPROVED			**APPROVED**		
Print	Signature	Date (Y-M-D)	Print	Signature	Date (Y-M-D)

	INSPECTION DATA SHEET (IDS) HEAT EXCHANGERS METALLIC	DATE (Y-M-D)	REV.
		PAGE 9 OF 9	

NOTES AND COMMENTS

PURCHASER			VENDOR		
A APPROVED			APPROVED		
Print	Signature	Date (Y-M-D)	Print	Signature	Date (Y-M-D)

Pipe Schedules/Data

In order to assist the inspector in determining the required thickness and diameter of pipe, the following tables based on ANSI/ASME B36.10[9] and B36.19[10] outline pipe size and schedule number. In addition, various other data useful to the inspector such as weight, cross-sectional area, moment of inertia, etc., are also included.

Nominal Pipe Size Inches	Outside Diam. Inches	Identification Steel Iron Pipe Size	Identification Steel Sched. No.	Identification Stainless Steel Sched. No.	Wall Thickness (t) Inches	Inside Diameter (d) Inches	Area of Metal Square Inches	Transverse Internal Area (a) Square Inches	Transverse Internal Area (A) Square Feet	Moment of Inertia (I) Inches[4]	Weight Pipe Pounds per foot	Weight Water Pounds per foot of pipe	External Surface Sq. Ft. per foot of pipe	Section Modulus $\left(2\frac{I}{OD}\right)$
		...	...	10S	.049	.307	.0548	.0740	.00051	.00088	.19	.032	.106	.00437
1/8	0.405	STD	40	40S	.068	.269	.0720	.0568	.00040	.00106	.24	.025	.106	.00523
		XS	80	80S	.095	.215	.0925	.0364	.00025	.00122	.31	.016	.106	.00602
		...	...	10S	.065	.410	.0970	.1320	.00091	.00279	.33	.057	.141	.01032
1/4	0.540	STD	40	40S	.088	.364	.1250	.1041	.00072	.00331	.42	.045	.141	.01227
		XS	80	80S	.119	.302	.1574	.0716	.00050	.00377	.54	.031	.141	.01395
		...	...	10S	.065	.545	.1246	.2333	.00162	.00586	.42	.101	.178	.01736
3/8	0.675	STD	40	40S	.091	.493	.1670	.1910	.00133	.00729	.57	.083	.178	.02160
		XS	80	80S	.126	.423	.2173	.1405	.00098	.00862	.74	.061	.178	.02554
		...	...	5S	.065	.710	.1583	.3959	.00275	.01197	.54	.172	.220	.02849
		...	...	10S	.083	.674	.1974	.3568	.00248	.01431	.67	.155	.220	.03407
		STD	40	40S	.109	.622	.2503	.3040	.00211	.01709	.85	.132	.220	.04069
1/2	0.840	XS	80	80S	.147	.546	.3200	.2340	.00163	.02008	1.09	.102	.220	.04780
		...	160	...	.187	.466	.3836	.1706	.00118	.02212	1.31	.074	.220	.05267
		XXS	...	...	.294	.252	.5043	.050	.00035	.02424	1.71	.022	.220	.05772
		...	...	5S	.065	.920	.2011	.6648	.00462	.02450	.69	.288	.275	.04667
		...	...	10S	.083	.884	.2521	.6138	.00426	.02969	.86	.266	.275	.05655
		STD	40	40S	.113	.824	.3326	.5330	.00371	.03704	1.13	.231	.275	.07055
3/4	1.050	XS	80	80S	.154	.742	.4335	.4330	.00300	.04479	1.47	.188	.275	.08531
		...	160	...	.219	.612	.5698	.2961	.00206	.05269	1.94	.128	.275	.10036
		XXS	...	...	.308	.434	.7180	.148	.00103	.05792	2.44	.064	.275	.11032
		...	...	5S	.065	1.185	.2553	1.1029	.00766	.04999	.87	.478	.344	.07603
		...	...	10S	.109	1.097	.4130	.9452	.00656	.07569	1.40	.409	.344	.11512
		STD	40	40S	.133	1.049	.4939	.8640	.00600	.08734	1.68	.375	.344	.1328
1	1.315	XS	80	80S	.179	.957	.6388	.7190	.00499	.1056	2.17	.312	.344	.1606
		...	160	...	.250	.815	.8365	.5217	.00362	.1251	2.84	.230	.344	.1903
		XXS	...	...	.358	.599	1.0760	.282	.00196	.1405	3.66	.122	.344	.2136
		...	...	5S	.065	1.530	.3257	1.839	.01277	.1038	1.11	.797	.435	.1250
		...	...	10S	.109	1.442	.4717	1.633	.01134	.1605	1.81	.708	.435	.1934
		STD	40	40S	.140	1.380	.6685	1.495	.01040	.1947	2.27	.649	.435	.2346
1¼	1.660	XS	80	80S	.191	1.278	.8815	1.283	.00891	.2418	3.00	.555	.435	.2913
		...	160	...	.250	1.160	1.1070	1.057	.00734	.2839	3.76	.458	.435	.3421
		XXS	...	...	.382	.896	1.534	.630	.00438	.3411	5.21	.273	.435	.4110
		...	...	5S	.065	1.770	.3747	2.461	.01709	.1579	1.28	1.066	.497	.1662
		...	...	10S	.109	1.682	.6133	2.222	.01543	.2468	2.09	.963	.497	.2598
		STD	40	40S	.145	1.610	.7995	2.036	.01414	.3099	2.72	.882	.497	.3262
1½	1.900	XS	80	80S	.200	1.500	1.068	1.767	.01225	.3912	3.63	.765	.497	.4118
		...	160	...	.281	1.338	1.429	1.406	.00976	.4824	4.86	.608	.497	.5078
		XXS	...	...	.400	1.100	1.885	.950	.00660	.5678	6.41	.42	.497	.5977
		...	...	5S	.065	2.245	.4717	3.958	.02749	.3149	1.61	1.72	.622	.2652
		...	...	10S	.109	2.157	.7760	3.654	.02538	.4992	2.64	1.58	.622	.4204
		STD	40	40S	.154	2.067	1.075	3.355	.02330	.6657	3.65	1.45	.622	.5606
2	2.375	XS	80	80S	.218	1.939	1.477	2.953	.02050	.8679	5.02	1.28	.622	.7309
		...	160	...	.344	1.687	2.190	2.241	.01556	1.162	7.46	.97	.622	.979
		XXS	...	...	.436	1.503	2.656	1.774	.01232	1.311	9.03	.77	.622	1.104
		...	...	5S	.083	2.709	.7280	5.764	.04002	.7100	2.48	2.50	.753	.4939
		...	...	10S	.120	2.635	1.039	5.453	.03787	.9873	3.53	2.36	.753	.6868
		STD	40	40S	.203	2.469	1.704	4.788	.03322	1.530	5.79	2.07	.753	1.064
2½	2.875	XS	80	80S	.276	2.323	2.254	4.238	.02942	1.924	7.66	1.87	.753	1.339
		...	160	...	.375	2.125	2.945	3.546	.02463	2.353	10.01	1.54	.753	1.638
		XXS	...	...	.552	1.771	4.028	2.464	.01710	2.871	13.69	1.07	.753	1.997
		...	...	5S	.083	3.334	.8910	8.730	.06063	1.301	3.03	3.78	.916	.7435
		...	...	10S	.120	3.260	1.274	8.347	.05796	1.822	4.33	3.62	.916	1.041
		STD	40	40S	.216	3.068	2.228	7.393	.05130	3.017	7.58	3.20	.916	1.724
3	3.500	XS	80	80S	.300	2.900	3.016	6.605	.04587	3.894	10.25	2.86	.916	2.225
		...	160	...	.438	2.624	4.205	5.408	.03755	5.032	14.32	2.35	.916	2.876
		XXS	...	...	.600	2.300	5.466	4.155	.02885	5.993	18.58	1.80	.916	3.424

Identification, wall thickness and weights are extracted from ANSI B36.10 and B36.19. The notations STD, XS, and XXS indicate Standard, Extra Strong, and Double Extra Strong pipe respectively.

Transverse internal area values listed in "square feet" also represent volume in cubic feet per foot of pipe length.

Nominal Pipe Size Inches	Outside Diam. Inches	Identification Steel Iron Pipe Size	Identification Steel Sched. No.	Identification Stainless Steel Sched. No.	Wall Thickness (t) Inches	Inside Diameter (d) Inches	Area of Metal Square Inches	Transverse Internal Area (a) Square Inches	Transverse Internal Area (A) Square Feet	Moment of Inertia (I) Inches^4	Weight Pipe Pounds per foot	Weight Water Pounds per foot of pipe	External Surface Sq. Ft. per foot of pipe	Section Modulus $\left(2\frac{I}{OD}\right)$
		...	...	5S	.083	3.834	1.021	11.545	.08017	1.960	3.48	5.00	1.047	.9799
		...	...	10S	.120	3.760	1.463	11.104	.07711	2.755	4.97	4.81	1.047	1.378
3½	4.000	STD	40	40S	.226	3.548	2.680	9.886	.06870	4.788	9.11	4.29	1.047	2.394
		XS	80	80S	.318	3.364	3.678	8.888	.06170	6.280	12.50	3.84	1.047	3.140
		...	...	5S	.083	4.334	1.152	14.75	.10245	2.810	3.92	6.39	1.178	1.249
		...	...	10S	.120	4.260	1.651	14.25	.09898	3.963	5.61	6.18	1.178	1.761
		STD	40	40S	.237	4.026	3.174	12.73	.08840	7.233	10.79	5.50	1.178	3.214
4	4.500	XXS	80	80S	.337	3.826	4.407	11.50	.07986	9.610	14.98	4.98	1.178	4.271
		...	120	...	.438	3.624	5.595	10.31	.0716	11.65	19.00	4.47	1.178	5.178
		...	160	...	.531	3.438	6.621	9.28	.0645	13.27	22.51	4.02	1.178	5.898
		XXS	...	...	.674	3.152	8.101	7.80	.0542	15.28	27.54	3.38	1.178	6.791
		...	...	5S	.109	5.345	1.868	22.44	.1558	6.947	6.36	9.72	1.456	2.498
		...	...	10S	.134	5.295	2.285	22.02	.1529	8.425	7.77	9.54	1.456	3.029
		STD	40	40S	.258	5.047	4.300	20.01	.1390	15.16	14.62	8.67	1.456	5.451
5	5.563	XS	80	80S	.375	4.813	6.112	18.19	.1263	20.67	20.78	7.88	1.456	7.431
		...	120	...	.500	4.563	7.953	16.35	.1136	25.73	27.04	7.09	1.456	9.250
		...	160	...	.625	4.313	9.696	14.61	.1015	30.03	32.96	6.33	1.456	10.796
		XXS	...	...	.750	4.063	11.340	12.97	.0901	33.63	38.55	5.61	1.456	12.090
		...	...	5S	.109	6.407	2.231	32.24	.2239	11.85	7.60	13.97	1.734	3.576
		...	...	10S	.134	6.357	2.733	31.74	.2204	14.40	9.29	13.75	1.734	4.346
		STD	40	40S	.280	6.065	5.581	28.89	.2006	28.14	18.97	12.51	1.734	8.496
6	6.625	XS	80	80S	.432	5.761	8.405	26.07	.1810	40.49	28.57	11.29	1.734	12.22
		...	120	...	.562	5.501	10.70	23.77	.1650	49.61	36.39	10.30	1.734	14.98
		...	160	...	.719	5.187	13.32	21.15	.1469	58.97	45.35	9.16	1.734	17.81
		XXS	...	...	.864	4.897	15.64	18.84	.1308	66.33	53.16	8.16	1.734	20.02
		...	...	5S	.109	8.407	2.916	55.51	.3855	26.44	9.93	24.06	2.258	6.131
		...	...	10S	.148	8.329	3.941	54.48	.3784	35.41	13.40	23.61	2.258	8.212
		...	20	...	.250	8.125	6.57	51.85	.3601	57.72	22.36	22.47	2.258	13.39
		...	30	...	.277	8.071	7.26	51.16	.3553	63.35	24.70	22.17	2.258	14.69
		STD	40	40S	.322	7.981	8.40	50.03	.3474	72.49	28.55	21.70	2.258	16.81
8	8.625	...	60	...	.406	7.813	10.48	47.94	.3329	88.73	35.64	20.77	2.258	20.58
		XS	80	80S	.500	7.625	12.76	45.66	.3171	105.7	43.39	19.78	2.258	24.51
		...	100	...	.594	7.437	14.96	43.46	.3018	121.3	50.95	18.83	2.258	28.14
		...	120	...	.719	7.187	17.84	40.59	.2819	140.5	60.71	17.59	2.258	32.58
		...	140	...	.812	7.001	19.93	38.50	.2673	153.7	67.76	16.68	2.258	35.65
		XXS	...	...	.875	6.875	21.30	37.12	.2578	162.0	72.42	16.10	2.258	37.56
		...	160	...	.906	6.813	21.97	36.46	.2532	165.9	74.69	15.80	2.258	38.48
		...	...	5S	.134	10.482	4.36	86.29	.5992	63.0	15.19	37.39	2.814	11.71
		...	...	10S	.165	10.420	5.49	85.28	.5922	76.9	18.65	36.95	2.814	14.30
		...	20	...	.250	10.250	8.24	82.52	.5731	113.7	28.04	35.76	2.814	21.15
		...	30	...	.307	10.136	10.07	80.69	.5603	137.4	34.24	34.96	2.814	25.57
		STD	40	40S	.365	10.020	11.90	78.86	.5475	160.7	40.48	34.20	2.814	29.90
10	10.750	XS	60	80S	.500	9.750	16.10	74.66	.5185	212.0	54.74	32.35	2.814	39.43
		...	80	...	.594	9.562	18.92	71.84	.4989	244.8	64.43	31.13	2.814	45.54
		...	100	...	.719	9.312	22.63	68.13	.4732	286.1	77.03	29.53	2.814	53.22
		...	120	...	.844	9.062	26.24	64.53	.4481	324.2	89.29	27.96	2.814	60.32
		XXS	140	...	1.000	8.750	30.63	60.13	.4176	367.8	104.13	26.06	2.814	68.43
		...	160	...	1.125	8.500	34.02	56.75	.3941	399.3	115.64	24.59	2.814	74.29
		...	...	5S	.156	12.438	6.17	121.50	.8438	122.4	20.98	52.65	3.338	19.2
		...	...	10S	.180	12.390	7.11	120.57	.8373	140.4	24.17	52.25	3.338	22.0
		...	20	...	.250	12.250	9.82	117.86	.8185	191.8	33.38	51.07	3.338	30.2
		...	30	...	.330	12.090	12.87	114.80	.7972	248.4	43.77	49.74	3.338	39.0
		STD	...	40S	.375	12.000	14.58	113.10	.7854	279.3	49.56	49.00	3.338	43.8
		...	40	...	.406	11.938	15.77	111.93	.7773	300.3	53.52	48.50	3.338	47.1
		XS	...	80S	.500	11.750	19.24	108.43	.7528	361.5	65.42	46.92	3.338	56.7
12	12.75	...	60	...	.562	11.626	21.52	106.16	.7372	400.4	73.15	46.00	3.338	62.8
		...	80	...	.688	11.374	26.03	101.64	.7058	475.1	88.63	44.04	3.338	74.6
		...	100	...	.844	11.062	1.53	96.14	.6677	561.6	107.32	41.66	3.338	88.1
		XXS	120	...	1.000	10.750	36.91	90.76	.6303	641.6	125.49	39.33	3.338	100.7
		...	140	...	1.125	10.500	41.08	86.59	.6013	700.5	139.67	37.52	3.338	109.9
		...	160	...	1.312	10.126	47.14	80.53	.5592	781.1	160.27	34.89	3.338	122.6

Identification, wall thickness and weights are extracted from ANSI B36.10 and B36.19. The notations STD, XS, and XXS indicate Standard, Extra Strong, and Double Extra Strong pipe respectively.

Transverse internal area values listed in "square feet" also represent volume in cubic feet per foot of pipe length.

Nominal Pipe Size Inches	Outside Diam. Inches	Identification: Steel: Iron Pipe Size	Identification: Steel: Sched. No.	Identification: Stainless Steel Sched. No.	Wall Thickness (t) Inches	Inside Diameter (d) Inches	Area of Metal Square Inches	Transverse Internal Area (a) Square Inches	Transverse Internal Area (A) Square Feet	Moment of Inertia (I) Inches[4]	Weight Pipe Pounds per foot	Weight Water Pounds per foot of pipe	External Surface Sq. Ft. per foot of pipe	Section Modulus $\left(\frac{2I}{OD}\right)$
14	14.00	...	...	5S	.156	13.688	6.78	147.15	1.0219	162.6	23.07	63.77	3.665	23.2
		...	...	10S	.188	13.624	8.16	145.78	1.0124	194.6	27.73	63.17	3.665	27.8
		...	10	...	.250	13.500	10.80	143.14	.9940	255.3	36.71	62.03	3.665	36.6
		...	20	...	.312	13.376	13.42	140.52	.9758	314.4	45.61	60.89	3.665	45.0
		STD	30	...	.375	13.250	16.05	137.88	.9575	372.8	54.57	59.75	3.665	53.2
		...	40	...	.438	13.124	18.66	135.28	.9394	429.1	63.44	58.64	3.665	61.3
		XS	...	...	.500	13.000	21.21	132.73	.9217	483.8	72.09	57.46	3.665	69.1
		...	60	...	.594	12.812	24.98	128.96	.8956	562.3	85.05	55.86	3.665	80.3
		...	80	...	.750	12.500	31.22	122.72	.8522	678.3	106.13	53.18	3.665	98.2
		...	100	...	.938	12.124	38.45	115.49	.8020	824.4	130.85	50.04	3.665	117.8
		...	120	...	1.094	11.812	44.32	109.62	.7612	929.6	150.79	47.45	3.665	132.8
		...	140	...	1.250	11.500	50.07	103.87	.7213	1027.0	170.28	45.01	3.665	146.8
		...	160	...	1.406	11.188	55.63	98.31	.6827	1117.0	189.11	42.60	3.665	159.6
16	16.00	...	...	5S	.165	15.670	8.21	192.85	1.3393	257.3	27.90	83.57	4.189	32.2
		...	...	10S	.188	15.624	9.34	191.72	1.3314	291.9	31.75	83.08	4.189	36.5
		...	10	...	.250	15.500	12.37	188.69	1.3103	383.7	42.05	81.74	4.189	48.0
		...	20	...	.312	15.376	15.38	185.69	1.2895	473.2	52.27	80.50	4.189	59.2
		STD	30	...	.375	15.250	18.41	182.65	1.2684	562.1	62.58	79.12	4.189	70.3
		XS	40	...	.500	15.000	24.35	176.72	1.2272	731.9	82.77	76.58	4.189	91.5
		...	60	...	.656	14.688	31.62	169.44	1.1766	932.4	107.50	73.42	4.189	116.6
		...	80	...	.844	14.312	40.14	160.92	1.1175	1155.8	136.61	69.73	4.189	144.5
		...	100	...	1.031	13.938	48.48	152.58	1.0596	1364.5	164.82	66.12	4.189	170.5
		...	120	...	1.219	13.562	56.56	144.50	1.0035	1555.8	192.43	62.62	4.189	194.5
		...	140	...	1.438	13.124	65.78	135.28	.9394	1760.3	223.64	58.64	4.189	220.0
		...	160	...	1.594	12.812	72.10	128.96	.8956	1893.5	245.25	55.83	4.189	236.7
18	18.00	...	...	5S	.165	17.670	9.25	245.22	1.7029	367.6	31.43	106.26	4.712	40.8
		...	...	10S	.188	17.624	10.52	243.95	1.6941	417.3	35.76	105.71	4.712	46.4
		...	10	...	.250	17.500	13.94	240.53	1.6703	549.1	47.39	104.21	4.712	61.1
		...	20	...	.312	17.376	17.34	237.13	1.6467	678.2	58.94	102.77	4.712	75.5
		STD	...	...	.375	17.250	20.76	233.71	1.6230	806.7	70.59	101.18	4.712	89.6
		...	30	...	.438	17.124	24.17	230.30	1.5990	930.3	82.15	99.84	4.712	103.4
		XS	...	...	.500	17.000	27.49	226.98	1.5763	1053.2	93.45	98.27	4.712	117.0
		...	40	...	.562	16.876	30.79	223.68	1.5533	1171.5	104.67	96.93	4.712	130.1
		...	60	...	.750	16.500	40.64	213.83	1.4849	1514.7	138.17	92.57	4.712	168.3
		...	80	...	.938	16.124	50.23	204.24	1.4183	1833.0	170.92	88.50	4.712	203.8
		...	100	...	1.156	15.688	61.17	193.30	1.3423	2180.0	207.96	83.76	4.712	242.3
		...	120	...	1.375	15.250	71.81	182.66	1.2684	2498.1	244.14	79.07	4.712	277.6
		...	140	...	1.562	14.876	80.66	173.80	1.2070	2749.0	274.22	75.32	4.712	305.5
		...	160	...	1.781	14.438	90.75	163.72	1.1369	3020.0	308.50	70.88	4.712	335.6
20	20.00	...	...	5S	.188	19.624	11.70	302.46	2.1004	574.2	39.78	131.06	5.236	57.4
		...	...	10S	.218	19.564	13.55	300.61	2.0876	662.8	46.06	130.27	5.236	66.3
		...	10	...	.250	19.500	15.51	298.65	2.0740	765.4	52.73	129.42	5.236	75.6
		STD	20	...	.375	19.250	23.12	290.04	2.0142	1113.0	78.60	125.67	5.236	111.3
		XS	30	...	.500	19.000	30.63	283.53	1.9690	1457.0	104.13	122.87	5.236	145.7
		...	40	...	.594	18.812	36.15	278.00	1.9305	1703.0	123.11	120.46	5.236	170.4
		...	60	...	.812	18.376	48.95	265.21	1.8417	2257.0	166.40	114.92	5.236	225.7
		...	80	...	1.031	17.938	61.44	252.72	1.7550	2772.0	208.87	109.51	5.236	277.1
		...	100	...	1.281	17.438	75.33	238.83	1.6585	3315.2	256.10	103.39	5.236	331.5
		...	120	...	1.500	17.000	87.18	226.98	1.5762	3754.0	296.37	98.35	5.236	375.5
		...	140	...	1.750	16.500	100.33	213.82	1.4849	4216.0	341.09	92.66	5.236	421.7
		...	160	...	1.969	16.062	111.49	202.67	1.4074	4585.5	379.17	87.74	5.236	458.5
22	22.00	...	...	5S	.188	21.624	12.88	367.25	2.5503	766.2	43.80	159.14	5.760	69.7
		...	...	10S	.218	21.564	14.92	365.21	2.5362	884.8	50.71	158.26	5.760	80.4
		...	10	...	.250	21.500	17.08	363.05	2.5212	1010.3	58.07	157.32	5.760	91.8
		STD	20	...	.375	21.250	25.48	354.66	2.4629	1489.7	86.61	153.68	5.760	135.4
		XS	30	...	.500	21.000	33.77	346.36	2.4053	1952.5	114.81	150.09	5.760	117.5
		...	60	...	.875	20.250	58.07	322.06	2.2365	3244.9	197.41	139.56	5.760	295.0
		...	80	...	1.125	19.75	73.78	306.35	2.1275	4030.4	250.81	132.76	5.760	366.4
		...	100	...	1.375	19.25	89.09	291.04	2.0211	4758.5	302.88	126.12	5.760	432.6
		...	120	...	1.625	18.75	104.02	276.12	1.9175	5432.0	353.61	119.65	5.760	493.8
		...	140	...	1.875	18.25	118.55	261.59	1.8166	6053.7	403.00	113.36	5.760	550.3
		...	160	...	2.125	17.75	132.68	247.45	1.7184	6626.4	451.06	107.23	5.760	602.4

Identification, wall thickness and weights are extracted from ANSI B36.10 and B36.19. The notations STD, XS, and XXS indicate Standard, Extra Strong, and Double Extra Strong pipe respectively.

Transverse internal area values listed in "square feet" also represent volume in cubic feet per foot of pipe length.

Nominal Pipe Size Inches	Outside Diam. Inches	Identification: Steel, Iron Pipe Size	Identification: Steel, Sched. No.	Identification: Stainless Steel Sched. No.	Wall Thickness (t) Inches	Inside Diameter (d) Inches	Area of Metal Square Inches	Transverse Internal Area (a) Square Inches	Transverse Internal Area (A) Square Feet	Moment of Inertia (I) Inches[4]	Weight Pipe Pounds per foot	Weight Water Pounds per foot of pipe	External Surface Sq. Ft. per foot of pipe	Section Modulus $\left(2\frac{I}{OD}\right)$
24	24.00	...	...	5S	.218	23.564	16.29	436.10	3.0285	1151.6	55.37	188.98	6.283	96.0
		...	10	10S	.250	23.500	18.65	433.74	3.0121	1315.4	63.41	187.95	6.283	109.6
		STD	20	...	.375	23.250	27.83	424.56	2.9483	1942.0	94.62	183.95	6.283	161.9
		XS	...	...	.500	23.000	36.91	415.48	2.8853	2549.5	125.49	179.87	6.283	212.5
		...	30	...	.562	22.876	41.39	411.00	2.8542	2843.0	140.68	178.09	6.283	237.0
		...	40	...	.688	22.624	50.31	402.07	2.7921	3421.3	171.29	174.23	6.283	285.1
		...	60	...	.969	22.062	70.04	382.35	2.6552	4652.8	238.35	165.52	6.283	387.7
		...	80	...	1.219	21.562	87.17	365.22	2.5362	5672.0	296.58	158.26	6.283	472.8
		...	100	...	1.531	20.938	108.07	344.32	2.3911	6849.9	367.39	149.06	6.283	570.8
		...	120	...	1.812	20.376	126.31	326.08	2.2645	7825.0	429.39	141.17	6.283	652.1
		...	140	...	2.062	19.876	142.11	310.28	2.1547	8625.0	483.12	134.45	6.283	718.9
		...	160	...	2.344	19.312	159.41	292.98	2.0346	9455.9	542.13	126.84	6.283	787.9
26	26.00	...	10	...	.312	25.376	25.18	505.75	3.5122	2077.2	85.60	219.16	6.806	159.8
		STD	...	...	.375	25.250	30.19	500.74	3.4774	2478.4	102.63	216.99	6.806	190.6
		XS	20	...	.500	25.000	40.06	490.87	3.4088	3257.0	136.17	212.71	6.806	250.5
28	28.00	...	10	...	.312	27.376	27.14	588.61	4.0876	2601.0	92.26	255.07	7.330	185.8
		STD	...	...	.375	27.250	32.54	583.21	4.0501	3105.1	110.64	252.73	7.330	221.8
		XS	20	...	.500	27.000	43.20	572.56	3.9761	4084.8	146.85	248.11	7.330	291.8
		...	30	...	.625	26.750	53.75	562.00	3.9028	5037.7	182.73	243.53	7.330	359.8
30	30.00	...	...	5S	.250	29.500	23.37	683.49	4.7465	2585.2	79.43	296.18	7.854	172.3
		...	10	10S	.312	29.376	29.10	677.76	4.7067	3206.3	98.93	293.70	7.854	213.8
		STD	...	...	.375	29.250	34.90	671.96	4.6664	3829.4	118.65	291.18	7.854	255.3
		XS	20	...	.500	29.000	46.34	660.52	4.5869	5042.2	157.53	286.22	7.854	336.1
		...	30	...	.625	28.750	57.68	649.18	4.5082	6224.0	196.08	281.31	7.854	414.9
32	32.00	...	10	...	.312	31.376	31.06	773.19	5.3694	3898.9	105.59	335.05	8.378	243.7
		STD	...	...	.375	31.250	37.26	766.99	5.3263	4658.5	126.66	332.36	8.378	291.2
		XS	20	...	.500	31.000	49.48	754.77	5.2414	6138.6	168.21	327.06	8.378	383.7
		...	30	...	.625	30.750	61.60	742.64	5.1572	7583.4	209.43	321.81	8.378	474.0
		...	40	...	.688	30.624	67.68	736.57	5.1151	8298.3	230.08	319.18	8.378	518.6
34	34.00	...	10	...	.344	33.312	36.37	871.55	6.0524	5150.5	123.65	377.67	8.901	303.0
		STD	...	...	.375	33.250	39.61	868.31	6.0299	5599.3	134.67	376.27	8.901	329.4
		XS	20	...	.500	33.000	52.62	855.30	5.9396	7383.5	178.89	370.63	8.901	434.3
		...	30	...	.625	32.750	65.53	842.39	5.8499	9127.6	222.78	365.03	8.901	536.9
		...	40	...	.688	32.624	72.00	835.92	5.8050	9991.6	244.77	362.23	8.901	587.7
36	36.00	...	10	...	.312	35.376	34.98	982.90	6.8257	5569.5	118.92	425.92	9.425	309.4
		STD	...	...	.375	35.250	41.97	975.91	6.7771	6658.9	142.68	422.89	9.425	369.9
		XS	20	...	.500	35.000	55.76	962.11	6.6813	8786.2	189.57	416.91	9.425	488.1
		...	30	...	.625	34.750	69.46	948.42	6.5862	10868.4	236.13	417.22	9.425	603.8
		...	40	...	.750	34.500	83.06	934.82	6.4918	12906.1	282.35	405.09	9.425	717.0

Identification, wall thickness and weights are extracted from ANSI B36.10 and B36.19. The notations STD, XS, and XXS indicate Standard, Extra Strong, and Double Extra Strong pipe respectively.

Transverse internal area values listed in "square feet" also represent volume in cubic feet per foot of pipe length.

Welding Symbols

ANSI/AWS A2.4[17] outlines welding symbols and nondestructive testing symbols used on fabrication drawings. This appendix outlines some of the common welding symbols and their usage. However, the inspector is referred to the American Welding Society standard for a complete outline of the various symbols.

Also included in this appendix are examples of dimensions for some of the common weld joint designs used in the fabrication of pressure equipment.

Finally, examples of a Welding Procedure Specification form (WPS) and a Welding Procedure Qualification Record Form (PQR), based on ASME Code guidelines, are provided for reference.

LOCATION OF ELEMENTS OF A WELDING SYMBOL

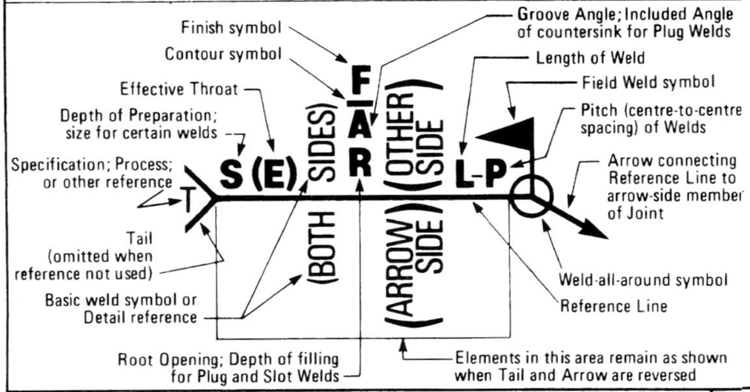

BASIC WELDING SYMBOLS and THEIR LOCATION SIGNIFICANCE

Location Significance	Fillet	Plug or Slot	Back or Backing	Stud	Arc Spot
ARROW SIDE			Groove Weld Symbol		
OTHER SIDE			Groove Weld Symbol	Not Used	
BOTH SIDES		Not Used	Not Used	Not Used	Not Used
No Arrow Side or Other Side Significance	Not Used	Not Used	Not Used	Not Used	Not Used

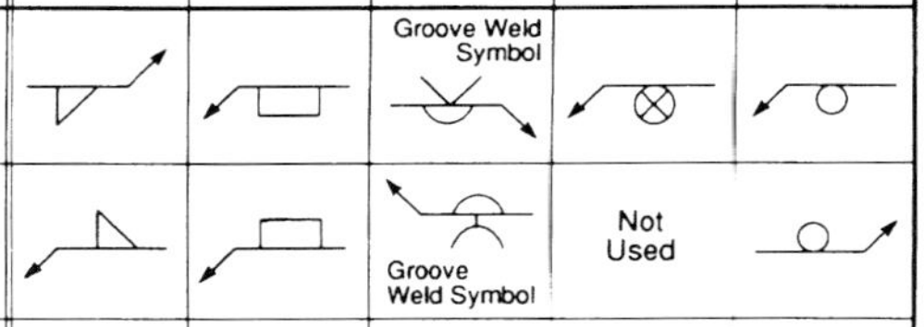

SUPPLEMENTARY SYMBOLS

Weld All-Around	Field Weld	Melt-Thru	BACKING, SPACER, OR CONSUMABLE INSERT	CONTOUR: Flush	CONTOUR: Convex	CONTOUR: Concave

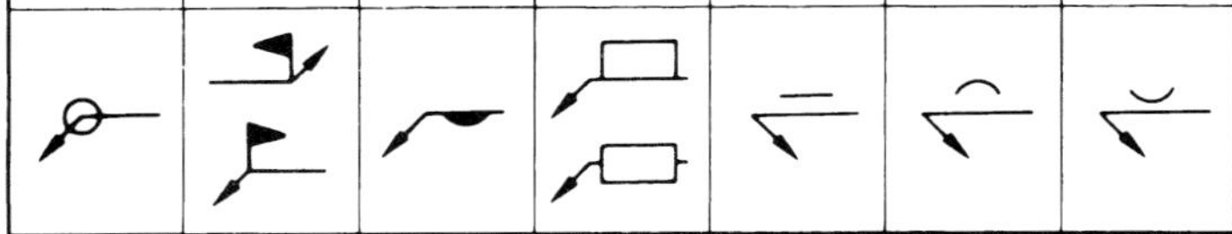

Where process abbreviations are to be included in the tail of the welding symbol, reference is made to Table A, Designation of Welding and Allied Processes by Letters, of AWS 2.4-79.

11 FLUSH CONTOUR SYMBOL

Flush contour symbol indicates face of weld to be made approximately flush.

When used without a surface finish symbol, it indicates weld without subsequent finishing

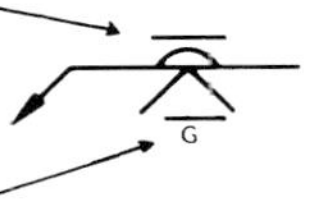

1 BASIC JOINTS BUTT JOINT

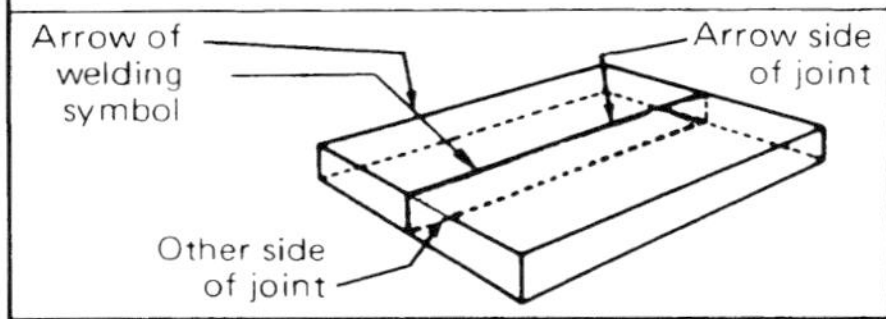

6 WELD-ALL-AROUND SYMBOL

Weld-all-around symbol indicates that weld extends completely around the joint

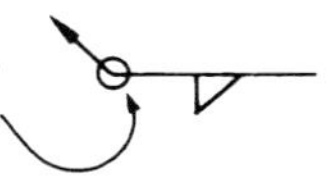

12 CONVEX CONTOUR SYMBOL

Finish symbol (user's standard) indicates method of obtaining specified contour but not degree of finish

Convex contour symbol indicates face of weld to be finished to convex contour by grinding

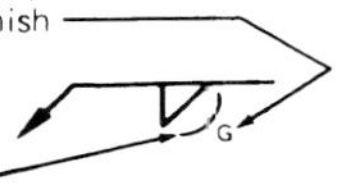

2 BASIC JOINTS T-JOINT

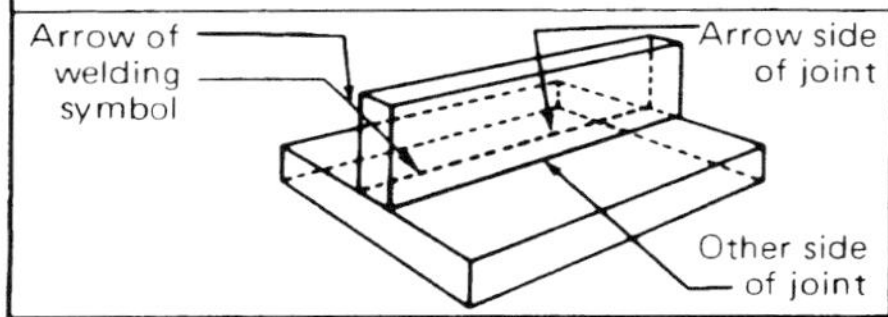

7 FIELD WELD SYMBOL

Flag points away from the arrow (W59 recommendation)

Field weld symbol indicates that weld is to be made at a place other than that of initial construction

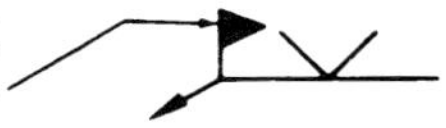

13 COMPLETE JOINT PENETRATION

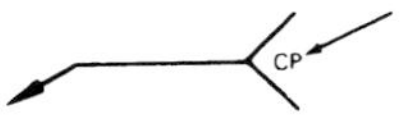

Indicates complete joint penetration groove weld regardless of type of weld and joint preparation

To be used on design drawing only (see Clause 4.1.1.1.4)

3 BASIC JOINTS CORNER JOINT

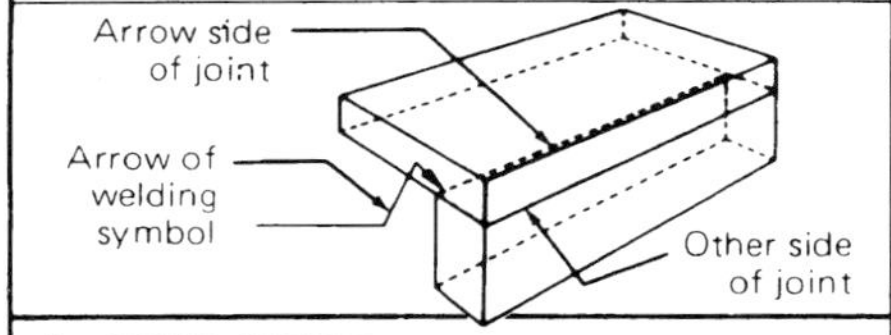

8 MELT-THRU SYMBOL

Any applicable weld symbol

Size (height) – Omission indicates no special height desired

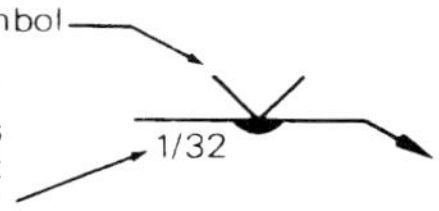

14 MULTIPLE REFERENCE LINES

First operation shown on reference line nearest arrow

Second operation, or supplementary data

Third operation, or test information

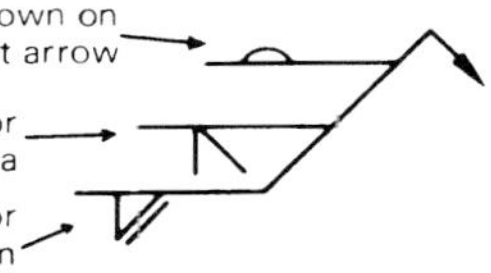

4 BASIC JOINTS LAP JOINT

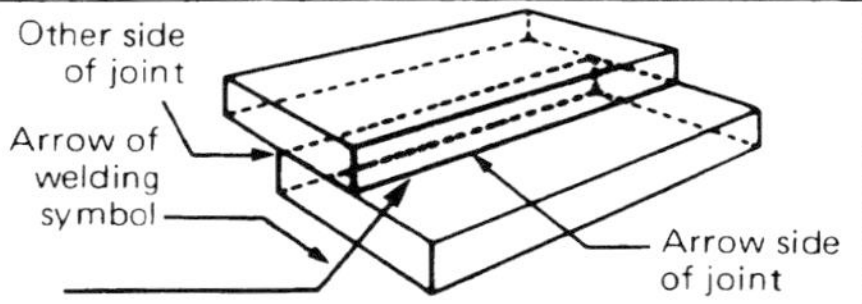

9 JOINT WITH BACKING

With groove-weld symbol

NOTES:
- Material (M) and dimensions of backing as specified.
- "R" indicates backing removed after welding.
- See Clause D1.6.

15 FINISHING OF WELDS

Finishing of welds other than cleaning is indicated by suitable contour and finish symbols. The following finishing symbols indicate the method, not the degree, of finish:

C – Chipping **R – Rolling**
G – Grinding **H – Hammering**
M – Machining

5 BASIC JOINTS EDGE JOINT

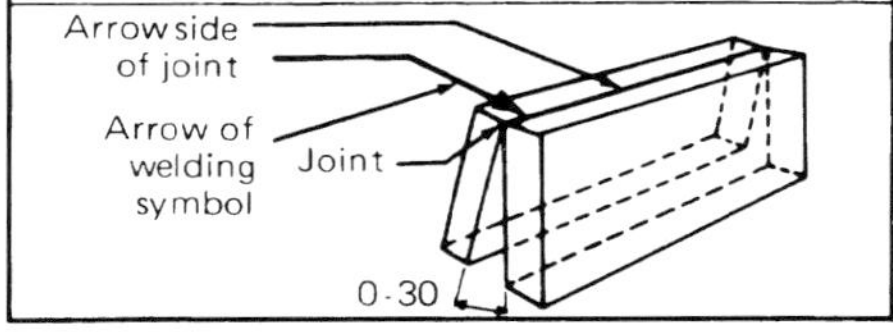

10 JOINT WITH SPACER

With modified groove-weld symbol

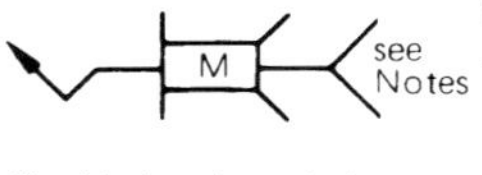

Double-bevel symbol

NOTES:
- Material (M) and dimensions of spacer as specified.
- See Clause D1.6.

16 ABBREVIATIONS

GTSM;– Gouge To Sound Metal
CP – Complete Penetration
PP – Partial Penetration

BASIC WELDING SYMBOLS AND THEIR LOCATION SIGNIFICANCE

Square Groove	V-Groove	Bevel-Groove	U-Groove	J-Groove	Flare-V-Groove	Flare-Bevel-Groove	Location Significance

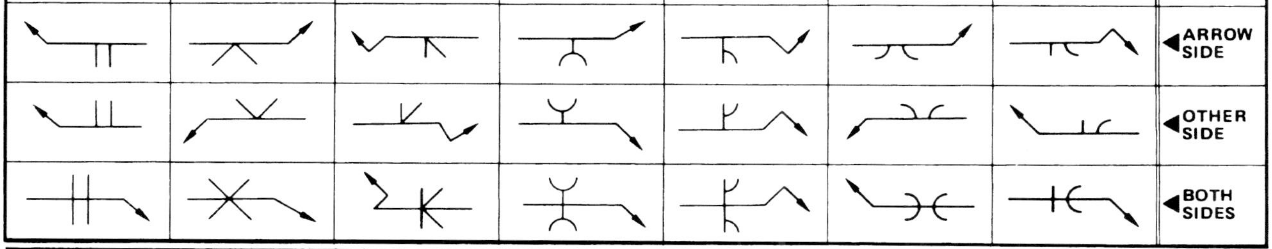

17 DOUBLE-FILLET WELDING SYMBOL

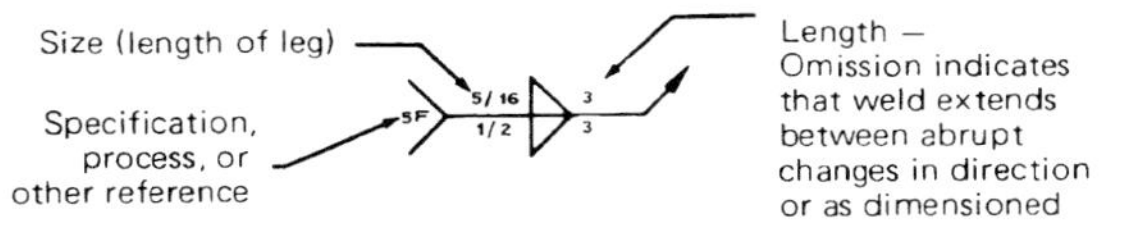

24 SINGLE-V-GROOVE WELDING SYMBOL

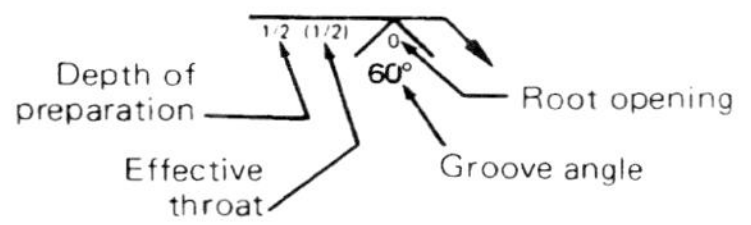

Omission of depth preparation and effective throat indicates complete joint penetration is required.

18 CHAIN INTERMITTENT FILLET WELDING SYMBOL

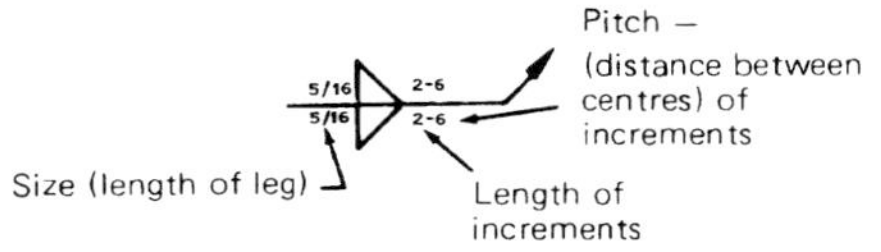

25 DOUBLE-BEVEL-GROOVE WELDING SYMBOL

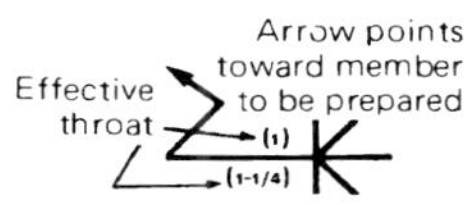

Omission of depth of preparation and effective throat indicates total depth of preparation — symmetrical in case of double-groove welds — equal to the thickness of the member and, hence, complete joint penetration groove weld is required

19 STAGGERED INTERMITTENT FILLET WELDING SYMBOL

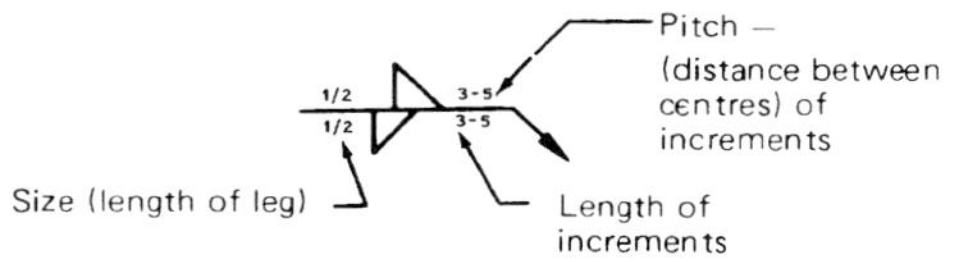

26 WELDING SYMBOL FOR GROOVES WELDED FROM BOTH SIDES WITH BACKGOUGING

Only depth of preparation is required in the welding symbol

GTSM signifies total effective throat is equal to the thickness of material. See Clause D1.2 of Appendix D.

20 PLUG WELDING SYMBOL

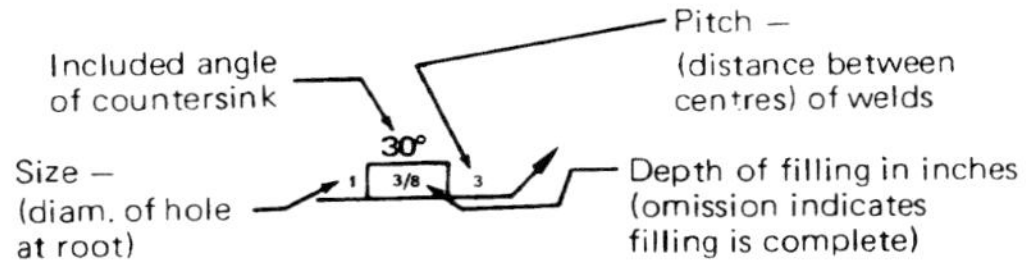

27 BACK-WELDING SYMBOLS

Any single-groove weld with a pass from the Other side with no special identification in the tail is considered a groove weld with a back weld.

21 SLOT WELDING SYMBOL

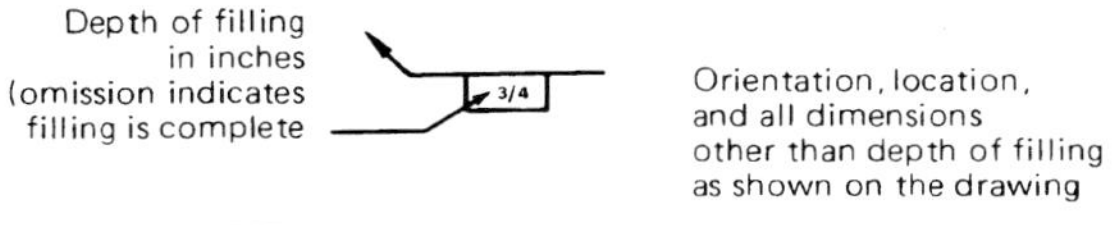

28 BACKING WELDING SYMBOLS

Any single-groove weld with a pass from one side identified in the tail by "backing weld" is considered to be a groove weld with a backing weld.

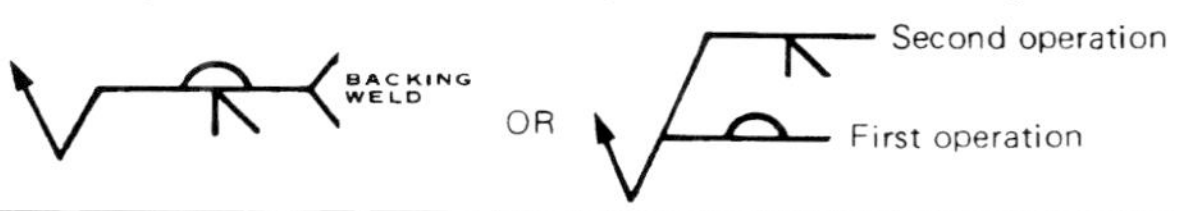

22 FLARE-V AND FLARE-BEVEL-GROOVE WELDING SYMBOLS

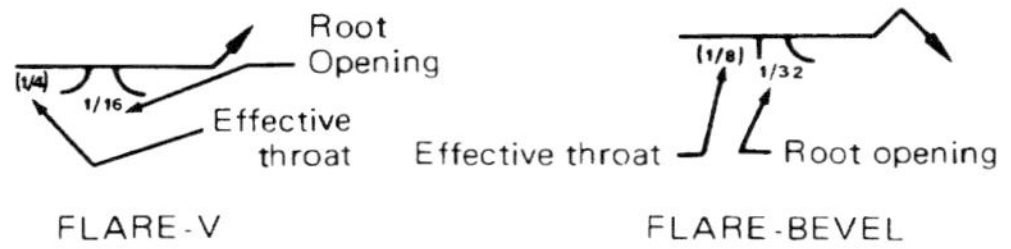

29 STUD WELDING SYMBOL

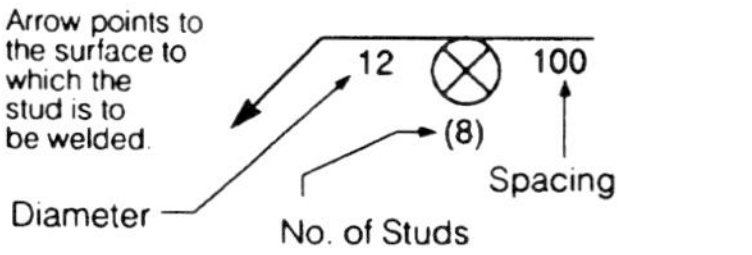

NOTES:
(1) - Stud Weld symbol is always below the reference line.
(2) - the location of the first and last Stud Weld in each single line shall be specified on the drawing.

23 SQUARE-GROOVE WELDING SYMBOL

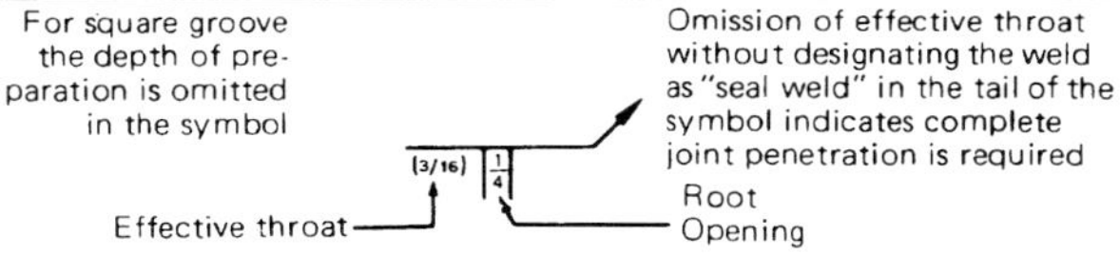

30 ARC SPOT WELDING SYMBOL

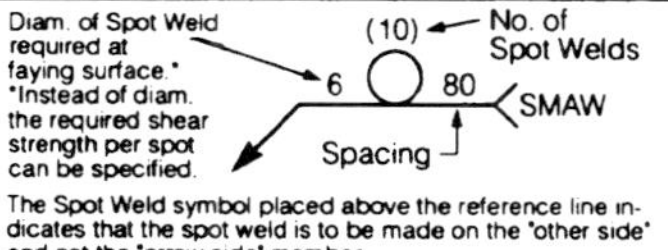

The Spot Weld symbol placed above the reference line indicates that the spot weld is to be made on the 'other side' and not the 'arrow side' member.

NOTES:
(1) - Arc Spot Weld symbols can be placed above or below the reference line, depending on whether welding is to be done on the 'other side' or the 'arrow side' member respectively.
(2) - The location of the first and last arc spot weld in each single line shall be specified on the drawing.

Typical Joints from a Shielded Metal Arc Welding Procedure

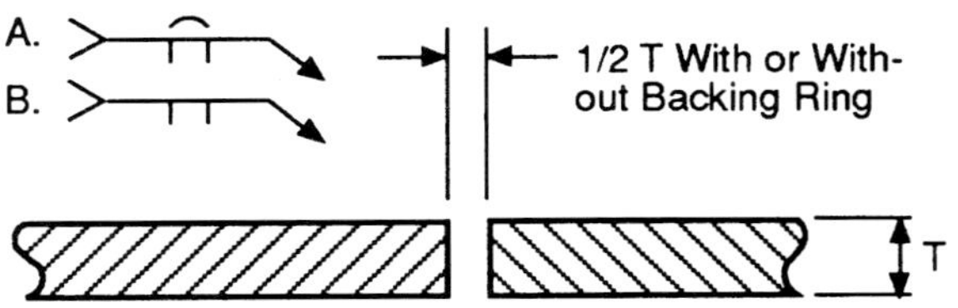

Straight Butt Joint
Wall thickness up to 3/8 in.

A. Up to 3/8 in. use straight butt joint if able to get inside to back chip and lay a weld pass
B. If inside inaccessible, straight butt joint to 3/16 in.

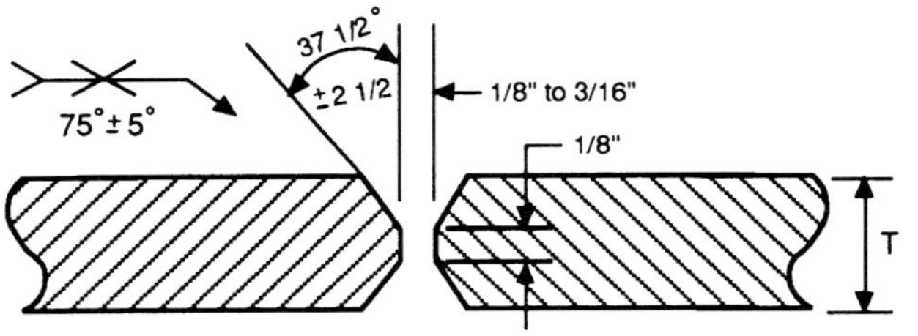

Double "V" Joint
Wall thickness 5/8 in. to 3/4 in. incl.
(Both sides accessible)

Back chip to sound weld metal and lay a pass

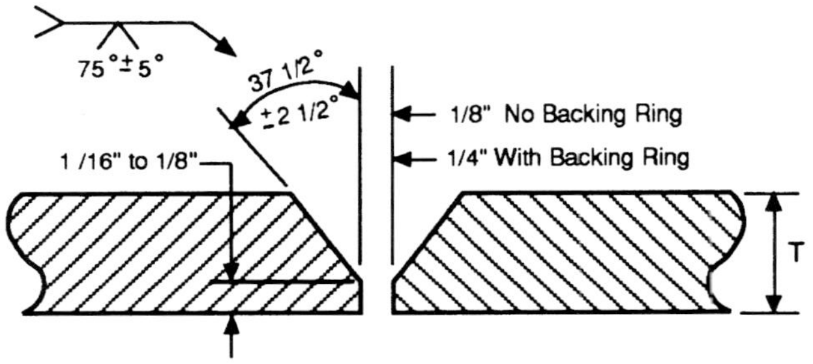

Single "V" Joint
Wall thickness 3/16 in. to 3/4 in. incl.
(One side accessible)

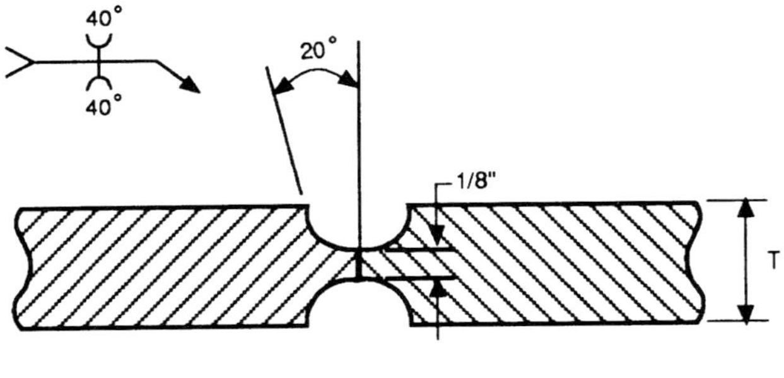

Double "U" Butt Joint
Wall thickness 5/8 in. to 3/4 in. incl.
Alternative for double "V" joint

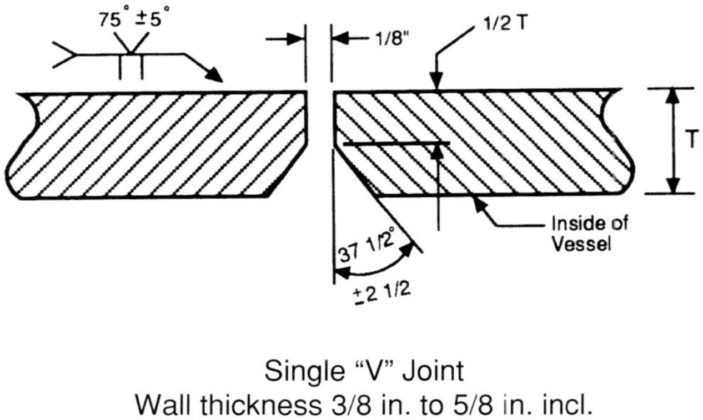

Single "V" Joint
Wall thickness 3/8 in. to 5/8 in. incl.
(Both sides accessible)
Back chip to sound weld metal on outside and lay final pass on outside.

Welding Procedure Specification (WPS)
Description de méthode de soudage (DMS)

② **Provincial Registration** / ***Enregistrement provincial***	WP / *MS*

① **Company/*Entreprise***	
Name/ *Nom*	
Address/ *Adresse*	
	Postal Code/ *Code postal*

③ **Procedure Identification/ *Désignation de la méthode***	
WPS No. / *DMS N°*	Revision No. / *Révision N°*
Code section / *Section du code*	
PQR No. / *RMS N°*	

④ **Welding Process(es)** / ***Procédé(s) de soudage***				
	1	Manual/ *Manuel*	Semi-auto.	Auto.
	2	Manual/ *Manuel*	Semi-auto.	Auto.
	3	Manual/ *Manuel*	Semi-auto.	Auto.

⑤ **Joints/*Joints*** (QW-402)			Sketch/ *Croquis*
Design/ *Genre*			
Backing / *Soutien*	Yes / *Oui*	No / *Non*	
Backing material / *Matériau de soutien*			
Other / *Autre*			

⑥ **Base Metals/*Métaux de base*** (QW-403)	P N° ______	Gr N° ______	to/ à	P N° ______	Gr N° ______
or Specifications type and grade / *ou Spécifications type et grade*			to/ à		
or Chemical Analysis and Mech Prop / *ou Analyse chimique et prop méc*			to/ à		
Thickness Range Qualified / *Gamme des épaisseurs qualifiées*	**Total**	1	2	3	
Other / *Autre*					

⑦ **Filler Metals/*Métaux d'apport*** (QW-404)				
1 AWS No./ *N°*	SFA	F-No	A-No	Dia
2 AWS No./ *N°*	SFA	F-No	A-No	Dia
3 AWS No./ *N°*	SFA	F-No	A-No	Dia
Flux Tradename / *Marque de commerce du flux*	Electrode-Flux Classification / *Classification du flux-électrode*			
Consumable Insert / *Pièce insérée consumable*	Other / *Autre*			

Positions/*Positions* (QW-405)		
Position(s) of Groove / *Position(s) du chanfrein*		
Welding Progression / *Progression du soudage*	Up / *Montant*	Down / *Descendant*
Position(s) of Fillet / *Position(s) de l'angle*		
Other / *Autre*		

⑩ **Postweld Heat Treatment/ *Traitement thermique postsoudage*** (QW-407)
Temperature Range / *Gamme de températures*
Time Range / *Gamme de durée*
Other / *Autre*

⑨ **Preheat/*Préchauffage*** (QW-406)	
Preheat Temperature / *Température de préchauffage*	Min
Interpass Temperature / *Température de l'interpasse*	Max
Preheat Maintenance / *Maintien du préchauffage*	
Other / *Autre*	

⑪ **Gas/*Gaz*** (QW-408)
Shielding Gas Composition / *Composition du gaz de protection*
Gas Flow Rate / *Débit du gaz*
Backing Gas Composition / *Composition du gaz de soutien*
Gas Flow Rate / *Débit du gaz*
Training Shielding Gas Composition / *Composition du gaz de protection trainant*

(12) **Electrical Characteristics and Technique/*Caractéristiques électriques et techniques*** (QW-409) - (QW-410)

Weld Layer(s) *Couche(s) de soudure*	Process *Procédé*	Filler Metal *Métal d'apport*		Current *Courant*				Travel Speed Range *Gamme de vitesse*
		Classification	Diameter *Diamètre*	Type *Type*	Polarity *Polarité*	Amp Range *Gamme d'ampérage*	Volt Range *Gamme de voltage*	

Tungsten Electrode Size and Type *Grosseur et type d'électrode de tungstène*		
Mode of Metal Transfer for GMAW *Mode de transfert du métal pour GMAW*		
String or Weave Bead *Cordon droit ou oscillant*		
Method of Back Gouging *Procédé de gougeage arrière*		
Initial and Interpass Cleaning (Brushing, Grinding, etc.) *Nettoyage initial et entre les passes (brossage, meulage, etc.)*		
Orifice or Gas Cup Size *Diamètre de la tuyère à gaz*		
Oscillation *Oscillation*	Width *Largeur*	Frequency *Fréquence*
Contact Tube to Work Distance *Distance entre tuyère et pièce*		
Multiple or Single Pass (per side) *Passe unique ou multiple (par côté)*		
Multiple or Single Electrode *Électrode unique ou multiple*		
Electrode Wire feed speed range *Gamme de vitesse d'alimentation de l'électrode*		
Other *Autre*		

(13) **Supplementary Comments/*Autres commentaires***

Procedure Qualification Record (PQR)
Résultat de méthode de soudage (RMS)

(Reference: ASME Section IX/*Référence: ASME, Section IX*)

① **Company/*Entreprise***	② **Provincial Registration** / ***Enregistrement provincial*** — WP / MS
Name/*Nom*	③ **Record Identification/*Désignation du résultat***
Address/*Adresse*	PQR No / *RMS N°*
	WPS No / *DMS N°*
Postal Code/*Code postal*	⑤
④ **Welding Process(es)/*Procédé(s) de soudage***	
1 Type	
2 Type	
3 Type	
⑥ **Base Metals/*Métaux de base*** (QW-403)	
P No. Gr No. to/*à* *P. N°* *Gr N°*	
Specification/*Spécification* Tensile/*Tension*	
	Groove Design used/*Croquis du joint utilisé*
Thickness / *Épaisseur* Diameter/*Diamètre*	⑩ **Postweld Heat Treatment/** ***Traitement thermique postsoudage*** (QW-407)
Other / *Autre*	Temperature / *Température*
⑦ **Filler Metals/*Métaux d'apport*** (QW-404)	Time / *Durée*
	Other / *Autre*

Classification (AWS)	SFA-No.	F - No.	A - No.

Other / *Autre*

⑧ **Positions/*Positions*** (QW-405)	⑪ **Gas/*Gaz*** (QW-408)
Position(s) of Groove / *Position(s) du joint*	Shielding Gas Composition / *Composition du gaz de protection*
Welding Progression / *Progression du soudage* — Up / *Montant* ☐ Down / *Descendant* ☐	Gas Flow Rate / *Débit du gaz*
Other / *Autre*	Backing Gas Composition / *Composition du gaz de soutien*
⑨ **Preheat/*Préchauffage*** (QW-406)	Gas Flow Rate / *Débit du gaz*
Preheat Temperature / *Température de préchauffage*	Other / *Autre*
Interpass Temperature / *Température de l'interpasse*	⑫ **Technique/*Technique*** (QW-410)
Other / *Autre*	String or Weave Bead / *Cordon droit ou oscillant*
	Oscillation / *Oscillation*
	Multipass or Single Pass (per side) / *Passe unique ou multiple (par côté)*
	Single or Multiple Electrode / *Électrode unique ou multiple*
	Other / *Autre*

⑬ **Electrical Characteristics and Technique/*Caractéristiques électriques et techniques*** (QW-409) + (QW-410)

Weld Layer(s) / *Couche(s) de soudure*	Process / *Procédé*	Filler Metal / *Métal d'apport*		Courrent / *Courant*				Travel Speed / *Vitesse*
		Classification	Diameter / *Diamètre*	Type / *Type*	Polarity / *Polarité*	Amp / *Ampérage*	Volt / *Voltage*	

(14) **Tansile Tests/*Essais à la traction*** — Reference: / *Référence*:

Specimen / *Éprouvette*	Dimensions / *Dimensions*: Width / *Largeur*	Dimensions / *Dimensions*: Thickness / *Épaisseur*	Area / *Surface*	Ultimate Total Load / *Charge limite*	Ultimate Unit Stress / *Contrainte limite*	Character and Location of Failure / *Nature et endroit de la rupture*

(15) **Guided Bend Tests/*Essais de pliage guidé*** — Reference: / *Référence*:

Specimen / *Éprouvette*	Face / *Face*	Root / *Racine*	Side / *Côté*	Result / *Résultat*	Specimen / *Éprouvette*	Face / *Face*	Root / *Racine*	Side / *Côté*	Result / *Résultat*

(16) **Toughness Tests/*Essais de résilience*** — Reference: / *Référence*:

Specimen / *Éprouvette*	Notch Location / *Endroit de l'entaille*	Notch Type / *Forme de l'entaille*	Specimen Temperature / *Température de l'éprouvette*	Impact Value / *Énergie de rupture*	Lateral Expansion / *Expansion latérale*: Shear / *% cisaillement*	Lateral Expansion / *Expansion latérale*: Mills / *Millièmes*	Drop Weight / *Essai en chute libre*: Break / *Cassé*	Drop Weight / *Essai en chute libre*: No Break / *Non cassé*

(17) **Fillet Weld Test/*Essai de soudure d'angle*** — Reference: / *Référence*:

Result - Satisfactory / *Résultat satisfaisant*	Yes / *Oui* ☐	No / *Non* ☐	Penetration Into Parent Metal / ***Pénétration dans métal de base***	Yes / *Oui* ☐	No / *Non* ☐
Macro - Results / *Résultats macroscopiques*					

(18) **Other Tests/*Autres essais*** — Reference: / *Référence*:

(19) Welder's Name / *Nom du soudeur* ______________________ Stamp No. / *N° de symbole* ____________

Tests conducted by / *Essais effectués par* ______________________ Laboratory Test No. / *Essai de laboratoire n°* ____________

We certify that the statements in this record are correct and that the test welds were prepared, welded and tested in accordance with the requirements of section IX of the ASME Code.

Nous certifions que les renseignements concernant ce résultat sont exacts et que les essais de soudage ont été préparés, soudés et exécutés conformément aux exigences de la section IX du Code ASME.

(20) Company Representative's Signature / *Signature du représentant de l'entreprise* ______________________ Date / *Date* ____________

(21) **Department Use Only/*Réservé au ministère***

Samples Welded on / *Pièces soudées le*	Inspector / *Inspecteur*
Record verified on / *Résultats vérifiés le*	Inspector / *Inspecteur*

Base Metal Groupings

To reduce the number of welding procedure qualifications required, base metals are grouped together according to similar characteristics such as composition, weldability, and mechanical properties. The groups are termed "P" or "P Numbers." The grouping is not intended to imply that base materials within a P Grouping may be indiscriminately substituted for the base material used in the qualification test without considering metallurgical properties, postweld heat treatment, mechanical properties, and other aspects that may affect the service life. Some division exists, as well, within the P Groupings producing sub-groupings of materials with similar properties.

Also, electrodes or filler metals are grouped together, again, to reduce the number of procedure qualifications. This grouping is based on usability characteristics, which determine the ability of welders to make satisfactory welds with a given filler metal. The grouping is termed the "F Number."

Grouping of filler metals and electrodes is not intended to imply that filler metals or base metals within specific groupings may be substituted for those used in the qualification testing without consideration of the properties of the finished product.

Finally, weld metal is classified as to the type of deposit made. This classification is termed the "A Number" and it indicates a general chemical composition for as-deposited weld metal. Deviation from the general composition is permitted, provided the physical properties are acceptable and the chemical composition is outlined in the procedure qualification.

P Groupings of Base Metals

P No.	Group No.	Material
1	1	Carbon steels with tensile strengths 40,000 to 75,000 psi (275 to 517 MPa)
1	2	C-Mn, C-Si, C-Mn-Si steels with tensile strengths 70,000 to 75,000 psi (483 to 517 MPa)
1	3	C-Mn-Si steels with tensile strengths to 80,000 psi (552 MPa)
3	1	Alloy steel with a chromium content of 0.75% maximum and generally with a total alloy content of 2% maximum; tensile strength 50,000 to 65,000 psi (345 to 448 MPa)
3	2	As for P3 Group I but with tensile strength 70,000 to 75,000 psi (483 to 517 Mpa)
3	3	Alloy steels containing Mn-Mo, Mn-Mo-Ni, Ni-Mo-Cr-V. Alloy contents are generally less than 2% but may be as high as 5% on specific alloys. Tensile strength is 80,000 to 90,000 psi minimum (552 to 621 MPa)
4	1	Alloy steels with chromium content between 0.75 and 2% with total alloy content generally not exceeding 3%. Tensile strength 60,000 to 75,000 psi minimum (414 to 517 MPa)

P No.	Group No.	Material
4	2	Alloys which may contain Cr, Cu, Ni, Al, Mo with the total alloy content less than 3% and a tensile strength of 60,000 psi (414 MPa) minimum
5	1	Alloy steel, generally Cr-Mo with the alloy content less than 5% and tensile strengths from 60,000 to 75,000 psi (414 to 517 MPa)
5	2	Alloy steel, generally Cr-Mo with alloy contents up to 10% and tensile strength of 60,000 to 100,000 psi (414 to 690 MPa) minimum
6	1	Martensitic, 11-13% Cr stainless steels with tensile strengths of 60,000 to 70,000 psi (414 to 483 MPa) minimum
6	2	Martensitic, 12-17% Cr stainless steels
6	3	Martensitic, Cr, Cr-Mo, Cr-Mo-Ni alloys
6	4	13% Cr-4% Ni alloy castings
7	1	Ferritic alloys 11-13% Cr
7	2	Ferritic Alloys 17-18% Cr
8	1	Austenitic stainless steels
8	3	17% Cr-4% Ni-6%Mn
		20% Cr-6% Ni-9% Mn
9A	1	Nickel steels, 2.5% maximum nickel content
9B	1	Nickel steels, 3.5% nickel content
10A	1	Mn-V steels
10B	2	Cr-V tubes
10C	3	Cr-Mn-Si plate
10D	4	20% Cr-1% Cu tubes
10B	5	26-27% Cr alloys
10F	6	Mn-Mo-V, Mn-Cr, Mo-V alloys
10G	7	36% Ni material
10H	8	18% Cr-5% Ni-3% Mo alloys
101	9	26% Cr-1% Mo alloys
11A	1	8-9% Ni alloys
11A	2	5% Ni-Mo alloys
11A	3	Castings Ni-Cr-Mo-V
11A	4	Mn-Mo, Mn-Mo-Ni material
11B	1-8	Various Cr-Mo-Si, Cr-Mo-Cu, Ni-Cr-Mo-V alloys
21		Aluminum alloys, tensile strengths 8,000 to 18,000 psi (55 to 124 MPa)
22		Aluminum alloys, tensile strengths 20,000 to 32,000 psi (138 to 221 MPa)
23		Aluminum alloys containing Si
25		Aluminum alloys, tensile strengths 33,000 to 45,000 psi (228 to 310 MPa)
31		Copper tensile strength 30,000 to 45,000 psi (207 to 310 MPa)
32		Copper alloys, tensile strength 40,000 to 50,000 psi (276 to 345 MPa)

P No. Group No.	Material
33	Copper - silicon alloys
34	Copper - nickel alloys
35	Bronzes
41	Nickel
42	Ni-Cu alloys
43	Ni-Cr-Fe, Ni-Cr-Mo-Cb, and Ni-Cr-Fe-Mo alloys
44	Ni-Mo, Ni-Cr-Mo alloys
45	Ni-Fe-Cr, Ni-Fe-Cr-Mo-Cu, Ni-Fe-Cr-Si, Cr-Ni-Fe-Mo-Cu-Cb alloys
5l	Unalloyed titanium, 35,000 to 50,000 psi (241 to 345 MPa) tensile strength
52	Unalloyed titanium, 65,000 psi (448 MPa) minimum tensile strength
61	Unalloyed zirconium

F NUMBERS

Grouping of Electrodes and Welding Rods for Qualification Steel and Steel Alloys

F. No.	ASME Specification Number	AWS Classification Number
l	SFA 5.1, 5.5	EXX 20,24, 27, 28
2	SFA 5.1, 5.5	EXX 12, 13, 14
3	SFA 5.1, 5.5	EXX 10, 11
4	SFA 5.1, 5.5	EXX 15, 16, 18
4	SFA 5.4, total alloy 6% or less	EXX 15, 16
4	SFA 5.4, total alloy more than 6%	EXX 15, 16
5	SFA 5.4 Cr-Ni	EXX 15, 16
6	SFA 5.2	RG XX
6	SFA 5.17	FXX-XXXX
6	SFA 5.9	ERXX
6	SFA 5.18	EXXXS-X, EXXU-X
6	SFA 5.20	EXXT-X
6	SFA 5.22	EXXXT-X
6	SFA 5.23	FXX-EXXX-X, FXX-ECXXX-X, FXX-EXXX-XN, FXX-ECXXX-XN

Aluminum and Aluminum-Based Alloys

F. No.	ASME Specification Number	AWS Classification Number
21	SFA 5.10	ER 1000, ER 1260
22	SFA 5.10	ER 5554, 5356, 5556, 5183, 5654
23	SFA 5.10	ER 4043
24	SFA 5.10	R-SC5lA, R-SG 70A

Copper and Copper-Based Alloys

F. No.	ASME Specification Number	AWS Classification Number
31	SFA 5.6, 5.7	RCu, ECu
32	SFA 5.6, 5.7	RCuSi-A, ECuSi
33	SFA 5.6, 5.7	RCuSn-A, ECuSn-A, ECuSn-C
34	SFA 5.6, 5.7	RCuNi, ECuNi
35	SFA 5.6, 5.7	RBCuZn-A, RCuZn-C
36	SFA 5.6, 5.7	ECuA1-A1, RCuA1-A2, ECuA1-A2, RCuA1-B, ECuA1-B

Nickel and Nickel-Based Alloys

F. No.	ASME Specification Number	AWS Classification Number
41	SFA 5.11	ENi-1
41	SFA 5.14	ENi-1
42	SFA 5.11	ENiCu-7
42	SFA 5.14	ERNiCu-7
43	SFA 5.11	ENiCrFe-1
43	SFA 5.11	ENiCrFe-2
43	SFA 5.11	ENiCrFe-3
43	SFA 5.11	ENiCrFe-4
43	SFA 5.11	ENiCrMo-2
43	SFA 5.11	ENiCr-3
43	SFA 5.14	ERNiCr-3
43	SFA 5.14	ERNiCrFe-5
43	SFA 5.14	ERNiCrFe-6
43	SFA 5.14	ERNiCrMo-2
43	SFA 5.14	ERNiCrMo-3

F. No.	ASME Specification Number	AWS Classification Number
44	SFA 5.11	ENiMo-1
44	SFA 5.11	ENiCrMo-4
44	SFA 5.11	ENiCrMo-5
44	SFA 5.14	ERNiMo-1
44	SFA 5.14	ERNiMo-2
44	SFA 5.14	ERNiCrMo-4
44	SFA 5.14	ERNiCrMo-5
45	SFA 5.11	ENiCrMo-1
45	SFA 5.14	ERNiCrMo-1
45	SFA 5.14	ERNiFeCr-1

Unalloyed Titanium

F. No.	ASME Specification Number	AWS Classification Number
51	SFA 5.16	ERTi-1, ERTi-2, ERTi-3, ER-Ti-4

A NUMBERS

Classification of Weld Metal Analysis for Procedure Qualifications

A No.	Deposit	Wt % C	Cr	Mo	Ni	Mn	Si
1	Mild Steel	0.15				1.60	1.00
2	Carbon-Moly	0.15	0.50 0.65	0.40-		1.60	1.00
3	Chrome (0.4-2%)-Moly	0.15	0.40-2.00	0.40-0.65		1.60	1.00
4	Chrome (2.0-6%)-Moly	0.15	2.00-6.00	0.40-1.50		1.60	2.00
5	Chrome (6-10.5%)-Moly	0.15	6.00-10.50	0.40-1.50		1.20	2.00
6	Chrome-Martensitic	0.15	11.00-15.00	0.70		2.00	1.00
7	Chrome-Ferritic	0.15	11.00-30.00	1.00		1.00	3.00
8	Chrome-Nickel	0.15	14.50-30.00	4.00	7.50-15.00	7.50-	1.00
9	Chrome-Nickel	0.30	25.00-30.00	4.00	15.00-37.00	2.50	1.00
10	Nickel to 4%	0.15		0.55	0.80-4.00	1.70	1.00
11	Nickel-Chrome-Moly	0.15	1.50	0.25-0.80	1.25-2.80	0.75-2.25	1.00

Operator Qualifications

The following pages outline the requirements for the qualifications of nondestructive testing operators according to the American Society for Nondestructive Testing (ASNT)[13] or to the Canadian Society for Nondestructive Testing, Inc. (CSNDT)[14] permitting the inspector to compare them with those of in-house training programs.

[13]The American Society for Nondestructive Testing, Columbus, OH.
[14]Canadian Society for Nondestructive Testing Inc., Mississauga, Ontario, Canada.

Table 1 Eligibility Requirements for Levels I and II

Education Level	Training Hours (Min.)		Work Time Experience Months (Min.)	
	Level I	Level II	Level I	Level II
a. Successful completion of at least two years of engineering science study at a university, college, or technical school	21	47	3	9
b. Secondary school graduation, diploma, or its equivalent	33	47	3	9
c. Elementary school graduation, or demonstrated proficiency or additional training	100	123	3	9

NOTES:

1. Training shall be as outlined in the Recommended Course Guidelines, Appendix A.

2. The work time experience factor in months is based on a nominal 40 h/week (175 h/month). When work is performed in excess of a 40 h/week, credit may be based on total hours. The candidate shall be responsible for providing evidence of work experience to the Certifying Agency.

3. For Level II certification, the work time experience shall consist of time as a certified Level II radiographer. If a person is being qualified directly to Level II with no time at Level I, the required experience shall consist of the sum of the times required for Level I and Level II, and the hours of training required for Level I and Level II in total shall apply.

4. Credit for work time experience may be gained simultaneously in two or more disciplines of nondestructive testing covered by CGSB standards on certification. The candidate must spend at least 25% of his work time on each discipline for which experience is bering claimed. Such credits shall be subject to approval by the Certifying Agency.

Table 2 Eligibility Requirements for Level III

	Work Time Experience (as certified Level II Industrial Radiographer) Years (Min.)
a. Graduate of a four-year accredited engineering or science college or university	1
b. Successful completion of at least two years of engineering science study at an accredited university, college, or technical school	2
c. Certified Level II Industrial Radiographer	4

NOTES:

Condition a. Experience in an assignment in industrial radiography comparable to that of a Level II industrial radiographer may be accepted.

If the college or university degree is issued in nondestructive testing, the one year's experience requirement may be reduced to six months.

Condition a.-c. The listed experience requirements may be partially replaced by up to 25% work time experience as a certified Level II operator in another method or methods covered by CGSB standards on certification.

Table 1 Eligibility Requirements for Levels I and II

Education Level	Training Hours (Min.)		Work Time Experience Months (Min.)	
	Level I	Level II	Level I	Level II
a. Successful completion of at least two years of engineering science study at a university, college, or technical school	24	40	3	9
b. Secondary school graduation, diploma, or its equivalent	40	40	3	9
c. Elementary school graduation, or demonstrated proficiency or additional training	40	80	3	9

NOTES:

1. Training shall be as outlined in the Recommended Course Guidelines, Appendix A.

2. The work time experience factor in months is based on a nominal 40 h/week (175 h/month). When work is performed in excess of a 40 h/week, credit may be based on total hours. The candidate shall be responsible for providing evidence of work experience to the Certifying Agency.

3. For Level II certification, the work time experience shall consist of time as a certified Level I ultrasonic operator. If a person is being qualified directly to Level II with no time at Level I, the required experience shall consist of the sum of the times required for Level I and Level II, and the hours of training required for Level I and Level II in total shall apply.

4. Credit for work time experience may be gained simultaneously in two or more disciplines of nondestructive testing covered by CGSB standards on certification. The candidate must spend at least 25% of his work time on each discipline for which experience is bering claimed. Such credits shall be subject to approval by the Certifying Agency.

Table 2 Eligibility Requirements for Level III

	Work Time Experience (as certified Level II Industrial Ultrasonics) Years (Min.)
a. Graduate of a four-year accredited engineering or science college or university	1
b. Successful completion of at least two years of engineering science study at an accredited university, college, or technical school	2
c. Certified Level II in Industrial Ultrasonics	4

NOTES:

Condition a. Experience in an assignment in industrial ultrasonics comparable to that of a Level II industrial ultrasonics may be accepted.

If the college or university degree is issued in nondestructive testing, the one year's experience requirement may be reduced to six months.

Condition a.-c. The listed experience requirements may be partially replaced by up to 25% work time experience as a certified Level II operator in another method or methods covered by CGSB standards on certification.

Table 1 Eligibility Requirements for Levels I and II

Education Level	Training Hours (Min.)		Work Time Experience Months (Min.)	
	Level I	Level II	Level I	Level II
a. Successful completion of at least two years of engineering science study at a university, college, or technical school	8	4	1	3
b. Secondary school graduation, diploma, or its equivalent	12	8	1	3
c. Elementary school graduation, or demonstrated proficiency or additional training	24	16	1	3

NOTES:

1. Training shall be as outlined in the Recommended Course Guidelines, Appendix A.

2. The work time experience factor in months is based on a nominal 40 h/week (175 h/month). When work is performed in excess of a 40 h/week, credit may be based on total hours. The candidate shall be responsible for providing evidence of work experience to the Certifying Agency.

3. For Level II certification, the work time experience shall consist of time as a certified Level I magnetic particle operator. If a person is being qualified directly to Level II with no time at Level I, the required experience shall consist of the sum of the times required for Level I and Level II, and the hours of training required for Level I and Level II in total shall apply.

4. Credit for work time experience may be gained simultaneously in two or more disciplines of nondestructive testing covered by CGSB standards on certification. The candidate must spend at least 25% of his work time on each discipline for which experience is bering claimed. Such credits shall be subject to approval by the Certifying Agency.

Table 2 Eligibility Requirements for Level III

	Work Time Experience (as certified Level II Magnetic Particle Testing) Years (Min.)
a. Graduate of a four-year accredited engineering or science college or university	1
b. Successful completion of at least two years of engineering science study at an accredited university, college, or technical school	2
c. Certified Level II in Magnetic Particle testing	4

NOTES:

Condition a. Experience in an assignment in magnetic particle testing comparable to that of a Level II magnetic particle operator may be accepted.

If the college or university degree is issued in nondestructive testing, the one year's experience requirement may be reduced to six months.

Condition a.-c. The listed experience requirements may be partially replaced by up to 25% work time experience as a certified Level II operator in another method or methods covered by CGSB standards on certification.

Table 1 Eligibility Requirements for Levels I and II

Education Level	Training Hours (Min.)		Work Time Experience Months (Min.)	
	Level I	Level II	Level I	Level II
a. Successful completion of at least two years of engineering science study at a university, college, or technical school	4	4	1	2
b. Secondary school graduation, diploma, or its equivalent	4	8	1	2
c. Elementary school graduation, or demonstrated proficiency or additional training	12	16	1	2

NOTES:

1. Training shall be as outlined in the Recommended Course Guidelines, Appendix A.

2. The work time experience factor in months is based on a nominal 40 h/week (175 h/month). When work is performed in excess of a 40 h/week, credit may be based on total hours. The candidate shall be responsible for providing evidence of work experience to the Certifying Agency.

3. For Level II certification, the work time experience shall consist of time as a certified Level I operator. If a person is being qualified directly to Level II with no time at Level I, the required experience shall consist of the sum of the times required for Level I and Level II, and the hours of training required for Level I and Level II in total shall apply.

4. Credit for work time experience may be gained simultaneously in two or more disciplines of nondestructive testing covered by CGSB standards on certification. The candidate must spend at least 25% of his work time on each discipline for which experience is bering claimed. Such credits shall be subject to approval by the Certifying Agency.

Table 2 Eligibility Requirements for Level III

	Work Time Experience (as certified Level II operator) Years (Min.)
a. Graduate of a four-year accredited engineering or science college or university	1
b. Successful completion of at least two years of engineering science study at an accredited university, college, or technical school	2
c. Certified Level II in Liquid Penetrants	4

NOTES:

Condition a. Experience in an assignment in liquid penetrant inspection comparable to that of a Level II liquid penetrant operator may be accepted.

If the college or university degree is issued in nondestructive testing, the one year's experience requirement may be reduced to six months.

Condition a.-c. The listed experience requirements may be partially replaced by up to 25% work time experience as a certified Level II operator in another method or methods covered by CGSB standards on certification.

Table 1 Eligibility Requirements for Levels I and II

Education Level	Training Hours (Min.)		Work Time Experience Months (Min.)	
	Level I	Level II	Level I	Level II
a. Successful completion of at least two years of engineering science study at a university, college, or technical school	12	11	1	9
b. Secondary school graduation, diploma, or its equivalent	17	13	1	9
c. Elementary school graduation, or demonstrated proficiency or additional training	55	32	1	9

NOTES:

1. Training shall be as outlined in the Recommended Course Guidelines, Appendix A.

2. The work time experience factor in months is based on a nominal 40 h/week (175 h/month). When work is performed in excess of a 40 h/week, credit may be based on total hours. The candidate shall be responsible for providing evidence of work experience to the Certifying Agency.

3. For Level II certification, the work time experience shall consist of time as a certified Level I operator. If a person is being qualified directly to Level II with no time at Level I, the required experience shall consist of the sum of the times required for Level I and Level II, and the hours of training required for Level I and Level II in total shall apply.

4. Credit for work time experience may be gained simultaneously in two or more disciplines of nondestructive testing covered by CGSB standards on certification. The candidate must spend at least 25% of his work time on each discipline for which experience is bering claimed. Such credits shall be subject to approval by the Certifying Agency.

Table 2 Eligibility Requirements for Level III

	Work Time Experience (as certified Level II eddy current operator) Years (Min.)
a. Graduate of a four-year accredited engineering or science college or university	1
b. Successful completion of at least two years of engineering science study at an accredited university, college, or technical school	2
c. Certified Level II - (Eddy Current Method)	4

NOTES:

Condition a. Experience in an assignment in eddy current inspection comparable to that of a Level II eddy current operator may be accepted.

If the college or university degree is issued in nondestructive testing, the one year's experience requirement may be reduced to six months.

Condition a.-c. The listed experience requirements may be partially replaced by up to 25% work time experience as a certified Level II operator in another method or methods covered by CGSB standards on certification.

Recommended Training and Experience Levels

Examination Method	Level	Technique	Completion with a passing grade of at least 2 years of engineering or science study in a university, college, or technical school — Training (hours)	High school graduation or equivalent — Training (hours)	Grammer school education, or demonstration proficiency, or additional training — Training (hours)	All eductational levels listed — Total work time experience (months) per level
RT	I		29	39	88	3
	II		35	40	80	9
MT	I		8	12	24	1
	II		4	8	16	3
UT	I		24	40	40	3
	II		40	40	80	9
PT	I		4	4	12	1
	II		4	8	16	2
ET	I		8	12	48	1
	II		8	8	24	9
NRT	I		20	28	56	6
	II		40	40	80	24
AE	I		40	60	80	6
	II		35	65	85	18
LT	I	BT	2	2	2	*
		PCMT	16	24	60	1 1/2
		HDLT	8	12	24	1 1/2
		MSLT	28	40	60	4
	II	BT	2	4	4	1/2
		PCMT	12	16	80	4
		HDLT	6	8	20	4
		MSLT	16	24	80	6

BT = Bubble Test PCMT = Pressure Change/Measurement Test HDLT = Halogen Diode Leak Test MSLT = Mass Spectrometer Leak Test * = 2 Hours

Notes:

(1) For Level II certification, the experience should consist of time at Level I or equivalent. If a person is being qualified directly to Level II with no time at Level I, the required experience should consist of the sum of the times required for Level I and Level II and the required training should consist of the sum of the hours required for Level I and Level II.

(2) Listed training guidelines may be prorated as described in the employer's written practice depending on the candidate's actual educational level.

(3) Initial experience may be gained simultaneously in two or more methods if:
 (a) The candidate spends a minimum of 25% of his work time on each method for which certification is sought, and
 (b) The remainder of his work time claimed as experience is spent in NDT-related activities as defined in the employer's written practice.

(4) Training should be outlined in the employer's written practice.

Bibliography

1. American Society of Mechanical Engineers (ASME). ASME Code Section III A, B, C, and Section VIII Division I, IX. New York, NY: ASME.
2. Chuse, Robert and Eber, Stephen, M. *Pressure Vessels: The ASME Code Simplified.* 6th ed. New York, NY: McGraw-Hill, 1984.
3. Chuse, Robert. *Pressure Vessels: The ASME Code Simplified.* 5th ed. New York, NY: McGraw-Hill, 1977.
4. Schlidt, Herbert. *ANSI Made Easy.* Berkeley, CA: Osborne McGraw-Hill, 1989.
5. American Petroleum Institute (API). *Guide for Inspection of Refinery Equipment.* Chapters I, II, III, IV, VI, VII, VIII, IX, X, XI. 3rd ed. Washington, DC: April 1982.
6. Kennedy, Clifford W., Andrews, Donald E. *Inspection and Gaging.* 5th ed. New York, NY: Industrial Press Inc., 1977.
7. Pilborough, Leslie. *Inspection of Industrial Plant: A Survey of Quality Assurance, Safety, and Standards.* Brookfield, IL: Gower Technical, 1989.
8. Society of Manufacturing Engineers. *Inspection Gaging and Testing.* 1st ed. Dearborn MI: 1984.
9. American National Standards Institute (ANSI). ANSI B36.I0, "Welded and Seamless Wrought Steel Pipe." New York, NY: ANSI.
10. Manufacturers Standardization Society of the Valve and Fittings Industry, Inc. (MSS). *Standard Practice SP-25, Standard Marking System for Valves, Fittings, Flanges and Unions.* Arlington, VA: MSS.
11. ANSI/American Welding Society, Inc. (AWS). *A2.4, Standard Symbols for Welding, Brazing and Nondestructive Examination.* Miami, FL: AWS.
12. American Cyanamid. *Nuts and Bolts*: "Identification and Proper Application of Fasteners." Westwego, LA: American Cyanamid, 1989.
13. Gillanders, John. *Pipe and Tube Bending Manual.* Houston, TX: Gulf Publishing Co., 1984.
14. Canadian Standards Association (CSA). *Steel Flanges.* Rexdale, Ontario, Canada: CSA, 1985.
15. Lee, Robert R. *Pocket Guide to Flanges, Fittings and Piping Data.* Houston, TX: Gulf Publishing Co., 1984.
16. Horger, Oscar J., ed. *ASME Handbook.* Vol. 1-4. New York, NY: McGraw-Hill, 1953-58.
17. Hill, Tom T. "What You Should Know About O.C.T.G. Inspections." *World Oil,* August, 1984, Part 1; September, 1984, Part 2; October, 1984, Part 3.
18. MacAdam, J.E., Vol. 53, *Inspection of Pressure Vessels in a Large Oil Refinery.* London: Institution of Mechanical Engineers, 1972.

19. Kletz, Trevor A. "The Prevention of Major Leaks—Better Inspection After Construction?" *Plant/Operations Progress*, Vol. 3, No. 1, Jan. 1984.

20. American Welding Society (AWS). *Welding Handbook*. Vol. 1-5, 7th ed. Miami, FL: AWS, 1976.

21. Tiivel, M., McGlynn, F.J., and Trickett, A.A. *Inspection Guidelines, Carbon Steel Sulfuric Acid Storage Tanks*. 1st ed. Toronto, Ontario, Canada: CIL, 1986.

22. American Welding Society (AWS). *Welding Inspection*. Miami, FL: AWS, 1980.

23. American Petroleum Institute (API). *Recommended Practice for Field Inspection of New Casing, Tubing, and Plain-End Drill Pipe*. Washington, DC: API, 1989.

Index

Notes

Notes

Notes

Notes